Also by

J. F. HORRABIN

o

AN ATLAS OF CURRENT AFFAIRS

"Just the right little book to have at the elbow when reading the day's news from the four corners of the earth."

—Charles A. Beard.

"A very important contribution to the understanding of dangerous problems which now confront the peoples of the earth. Admirable for young and old."

—James Harvey Robinson.

"A remarkable piece of work, quite indispensable for the understanding of contemporary politics."

—Harold J. Laski.

"A book which will be of great assistance to that ever-growing category of people who realise that 'foreign affairs' are matters of life or death to them, and are determined to find out what they are all about."

—Vernon Bartlett.

PUBLISHED BY ALFRED A. KNOPF

AN ATLAS OF EUROPEAN HISTORY

HISTORY

from the 2nd *to the* 20th *century*

AN ATLAS

of

EUROPEAN

HISTORY

from the 2nd *to the* 20th *century*

by

J. F. HORRABIN

NEW YORK · ALFRED · A · KNOPF

1935

PREFACE

Mr. Chesterton, expatiating recently on the importance of taking long views, wisely remarked that they must be views to the past as well as to the future. It was while working on *An Atlas of Current Affairs* that I realised how essential a background of historical knowledge was to the right understanding of present-day problems. Hence this book.

The maps in this volume have not been designed to illustrate "history for history's sake"—much less at covering completely the million-and-one facts of European development. I have aimed at bringing out those facts which appeared to me to have some more or less direct bearing on the Europe in which we live—and tremble—to-day. If it be objected that the facts selected are too exclusively those relating to political and territorial changes I can only answer, first, that it is difficult, if not impossible, effectively to illustrate the history of intellectual development by means of maps; and, second, that the historical background most relevant to the political and territorial maladjustments which are possible causes of war to-day is precisely an account of how those particular problems came to be what they are.

The first section (of 28 maps) tries to tell, in the baldest of outlines, the main European story from the break-up of the Roman Empire down to the Great War. How much has had perforce to be omitted from

such an outline I know only too well; but the attempt
to summarise certain essentials of the story in a short
sequence seemed worth while. The remaining 42
maps, which deal with various nation-states sepa-
rately, fill in some of the gaps in the first series. The
contents list opposite indicates the grouping and
sequence of the maps.

From the many excellent and learned historical
works which I have consulted for my facts I should
like to pick out three which I have found most
stimulating and suggestive: the historical section of
Mr. and Mrs. Cole's *Intelligent Man's Review of
Europe Today,* Mr. Frank J. Adkins' *How Europe
Grew,* and the late Ellen Churchill Semple's *In-
fluences of Geographic Environment.* But even so
small a work as this necessitates a vast deal of pains-
taking listing and collating of facts and dates and
frontier-lines from a score of sources; and for her
invaluable assistance in this connection, as well as
for many helpful suggestions, I have again to thank
Margaret McWilliams.

J. F. H.

Hampstead,
March 1935

CONTENTS

ix

AN ATLAS OF EUROPEAN HISTORY

HISTORY

from the 2nd *to the* 20th *century*

MAP 1

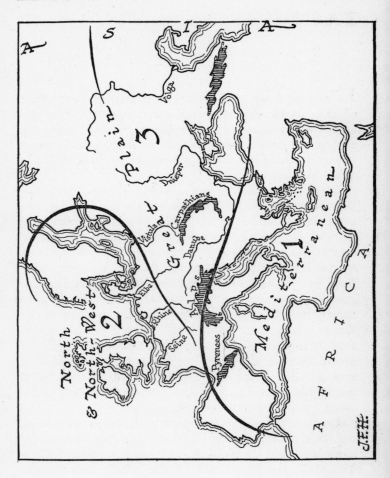

2

THE GEOGRAPHICAL
FACTORS

In the history of Europe, as in that of every part of the world, certain geographical factors have affected every stage of development. Let us begin by noting the broad geographical divisions which correspond quite definitely to particular periods of European history.

That history began in the Mediterranean area, the belt of sheltered inland seas which divides the two great land masses of the Old World, Eurasia and Africa. Mediterranean civilisation, in the course of centuries, spread northwards and north-westwards to the islands and peninsulas of the Atlantic coast-line, and from these to the countries bordering the great northern inland sea, the Baltic. Later still, and largely from the Baltic, it spread to the peoples of the Great Plain, the "mainland" proper which makes Europe a part of the Asian land mass—although the movements of the peoples of the Plain had indeed reacted earlier on the course of history, when, as barbarian nomads, they had forced their way westward into the more civilised areas.

Certain lands are, of course, transitional areas: e.g. Spain and France, linking the Mediterranean zone with the north and north-west; or the Balkan Peninsula and the Danube Valley, lying between the Mediterranean and the Great Plain.

MAP 2

Huns

Goths

Germans

Dacia

Mesopotamia

Euphrates

Tigris

Egypt

Danube

Elbe

Rhine

Alexandria

BRITAIN

GAUL

ALPS

Rome

Po

Numidia

SPAIN

Mauretania

J.F.H.

2nd century

4

THE ROMAN EMPIRE
AT ITS GREATEST
EXTENT

CRETANS, Egyptians, Phœnicians, Greeks, had established civilisation in the Mediterranean area; and the two last-named had carried it almost to the farthest shores of the inland sea. It remained for Rome to weld all the coastlands and islands into a political unit. Note that at this period the southern (African) coast of the Mediterranean was an integral part of the civilised world; the barrier between Europe and Africa was not the sea, but the great desert lying to the south of the coastlands.

The Romans brought under their control not only the lands actually bordering the Mediterranean, but extended their rule eastwards into Asia and—much more importantly from the point of view of the spread of civilisation—north-westwards to Gaul and Britain and the Rhine. The reigns of the emperors Trajan and Hadrian (end of the first century and beginning of the second century of our era) saw the empire of Rome at its peak.

5

MAP 3

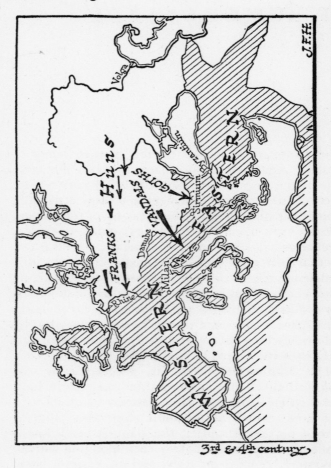

Volga

HUNS →

FRANKS

VANDALS *GOTHS* →

Danube

Sarmatians

Sirmium

Milan

Rhine

Rome

WESTERN

EASTERN

J.F.H.

3rd & 4th century

6

THE ·BREAK-UP
OF ROME

THE GRADUAL drying-up of the pasture lands of
Western Asia which took place at about the same
time as the beginning of the Christian era was the
direct cause of a vast movement and displacement of
peoples which resulted in the break-up of the Roman
power. After Hadrian, Rome expanded no more.
Thenceforth all her energies (and they, for various
internal reasons, steadily declined) were expended
on defending her northern and eastern frontiers from
attacks by barbarian races coming from the unknown
world outside her borders. There was a great southward drive of Germanic peoples from Scandinavia
and the Baltic, and a westward movement of the
peoples of the Great Plain.

The Mediterranean Sea is, geographically, obviously divided into a western and an eastern half;
and the fierce thrust of the barbarians, concentrated
to a large extent on the most vulnerable point of
Rome's northern frontier, the Danube bend, finally
broke the Empire into two corresponding halves:
a Western, centred on Italy; and an Eastern, based
on the city of Byzantium (later Constantinople).

MAP 4

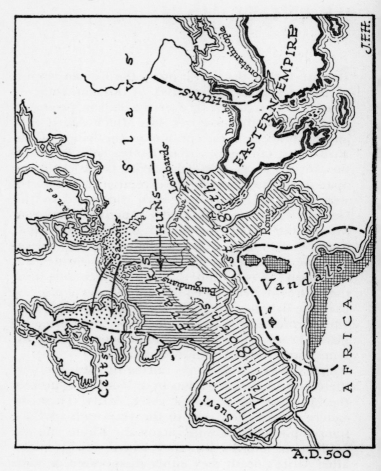

A.D. 500

8

THE BARBARIAN
"KINGDOMS"

By the end of the fifth century the Roman empire
of the west had disappeared. A Roman emperor
ruled in Constantinople over the Balkan peninsula
up to the Danube, and over Asia Minor. All western
and north-western Europe was in the hands of the
barbarians. Ostrogoths (East Goths) held Italy and
the Adriatic coast. Visigoths (West Goths) held the
Spanish peninsula, with the exception of the north-
western corner, which had been conquered by the
German tribe of the Suevi. Another Germanic
people, the Franks, held what is now France and
much of Germany. The northern coast of Africa,
with the islands of Corsica and Sardinia, formed the
"kingdom" of the Vandals. Angles and Saxons from
the coastlands between Rhine and Elbe were pushing
back the Britons in Britain. Only the Pope of Rome,
who at this time was breaking free from any alle-
giance to the Emperor at Constantinople, stood for
some sort of western European unity.

Into this chaotic Europe, during the fifth century,
there came a great invasion by a Mongolian people,
the Huns, under their war chief, Attila. These invaders pushed their way to the very walls of Constantinople, and westwards across the Rhine into the
land of the Franks. But they settled nowhere, and
after Attila's death they either disappeared or were
absorbed into existing European populations.

9

MAP 5

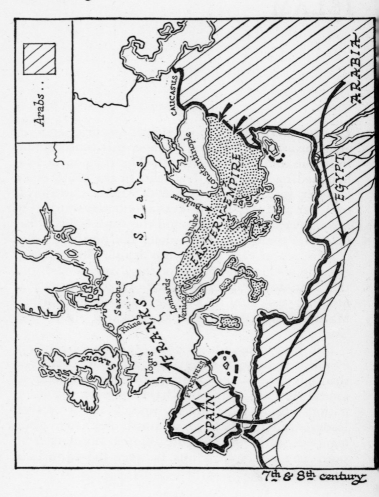

Arabs..

CAUCASUS

Constantinople

EASTERN EMPIRE

Bulgars

Danube

ARABIA

EGYPT

S l a v s

Saxons

Lombards

Venice

Rhine

Saxons

Dane

Tours

FRANKS

PYRENEES

SPAIN

7th & 8th century

10

THE COMING OF
ISLAM

IN THE SEVENTH and early eighth centuries another alien invasion threatened Europe, this time coming from the south and south-east. The religion of Mohammed had, within a few years of the death of its founder, spread from the oases of Arabia, northwards over Syria, Armenia and Persia, and westwards into Egypt. Another few years, and the Arab Caliphs had extended their rule right along the northern coast of Africa, and had crossed into Spain. Their all-conquering warriors pushed over the Pyrenees and into the Frankish kingdom beyond. In 732, exactly a hundred years after the death of Mohammed, Charles Martel, ruler of the Franks, defeated them near Tours.

But they consolidated their rule in Spain, and established there a civilisation far more advanced than any other then existing north of the Mediterranean. It was a civilisation based on extensive international trade—Mohammed himself was a caravan leader; and the Caliphates, until they, too, disintegrated under the pressure of new invaders from north and east, were a link between the cultures of Asia and Europe.

MAP 6

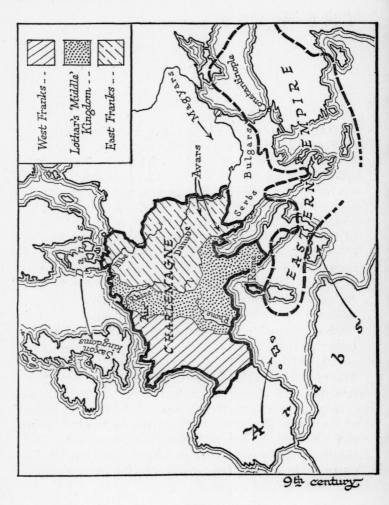

9th century

12

A NEW EMPIRE OF
THE WEST

CHARLES MARTEL (see previous map) had built up
a Frankish kingdom which extended from the
Pyrenees to the lands between Rhine and Elbe. His
grandson, Charles the Great (Charlemagne) pushed
his frontiers still further eastward, and, on the other
side of his kingdom, forced back the Arabs some
small distance, establishing a "Spanish March" south
of the Pyrenees. He and the Emperor of the East at
Constantinople shared the greater part of Europe
between them, with barbarian Danes and Norsemen
to the north, Slav peoples to the east (Serbs and
Bulgars had already established themselves in the
Balkan peninsula), and Arabs to the south.

Charlemagne conceived the idea of reviving the
imperial title, and, in return for the protection he
had given to the Pope, was crowned in Rome as
Emperor of the West on Christmas Day, A.D. 800.
But after his death his domains fell apart, and its
three divisions form the nucleus of the modern states
of France on the west, and Germany on the east; with
the "Middle Kingdom"—Lorraine, Burgundy, and
Northern Italy—the zone in which the frontier between France and Germany was to fluctuate for over
a thousand years to come.

MAP 7

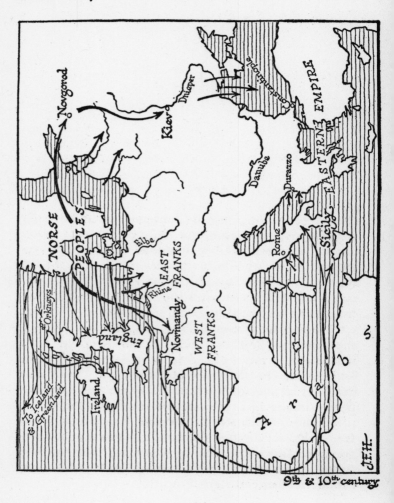

Novgorod

Kiev

Dnieper

Constantinople

EASTERN EMPIRE

Danube

Durazzo

NORSE

PEOPLES

Rome

Sicily

Elbe

EAST
FRANKS

Rhine

Normandy

WEST
FRANKS

Orkneys

England

A   r   a   b   s

To Iceland
& Greenland

Ireland

J.F.H.

9th & 10th century

14

THE NORSEMEN

Nᴏʀᴛʜ of the lands which Charlemagne had partially consolidated, the Scandinavian peoples still stood outside the pale of Christendom, as they had remained beyond the farthest frontiers of Rome. Theirs was a fighting and sea-faring civilisation, and in the ninth and tenth centuries they proved their vigour by raiding and settling west, south and east. Danes established themselves in Saxon England and along the coasts of Scotland and Ireland. Kindred races settled in the valley of the Seine in northern France, establishing the state of Normandy; and sailed into the Mediterranean, where they challenged the Arab power and settled in Sicily and southern Italy.

Another branch struck eastward, founding in what is now Russia the first centres of civilisation at Novgorod in the north, and at Kiev, on the Dnieper; thence pushing still further south and establishing contact with the Eastern Empire.

MAP 8

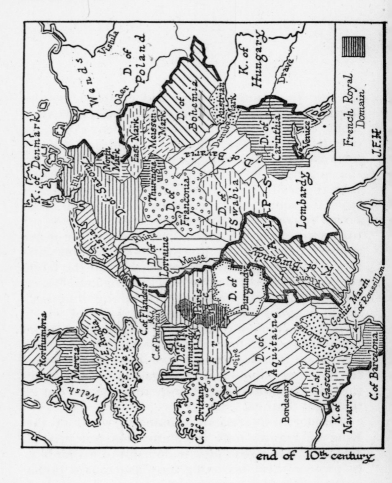

K. of Denmark

W e n d s

Vistula

Oder D. of Poland

K. of Hungary

Drave

Elbe

North Mark

East Mark

Saale

D. of Saxony

Thuringia

Meissen Mark

D. of Bohemia

Austrian Mark

Danube

K. of Carinthia

Venice

Frisia

Weser

D. of Franconia

D. of Bavaria

D. of Swabia

Rhine

D. of Lorraine

Meuse

A L P S

Lombardy

Rhône

K. of Burgundy

C. of Flanders

C. of Ponthieu

Paris

D. of Burgundy

Gothic March

C. of Roussillon

Northumbria

Welsh

Mercia

E. Anglia

Wessex

D. of Normandy

F r a n c e

Loire

D. of Aquitaine

D. of Toulouse

C. of Barcelona

C. of Brittany

Bordeaux

D. of Gascony

K. of Navarre

French Royal Domain

J.F.H.

end of 10th century

16

FEUDALISM

During the centuries which had elapsed since the break-up of Roman power there had been only spasmodic attempts at centralised government extending over large areas. The machinery of political government had to grow again on a smaller scale. Moreover, the universal threat of invasion by the Norsemen from the north, the Slavs from the east, and the Arabs from the south, made it inevitable that local rule, taking responsibility for defence, should develop all over north-western Europe. Thus grew up the social system we call Feudalism: a system based upon military over-lordship and land ownership. In the tenth century the states which we now call England, France and Germany were actually divided up into earldoms, counties and dukedoms, the lords of which owned the nominal suzerainty of a king or emperor, but were to all intents and purposes themselves independent rulers. Some idea of the extent of this territorial sub-division may be gained by a study of the map of north-western Europe at about the end of the tenth century.

William of Normandy, himself a vassal of the King of France, took care, after his conquest of England, so to subdivide the domains of his lords as to prevent them becoming too dangerous a threat to the power of the crown.

MAP 9

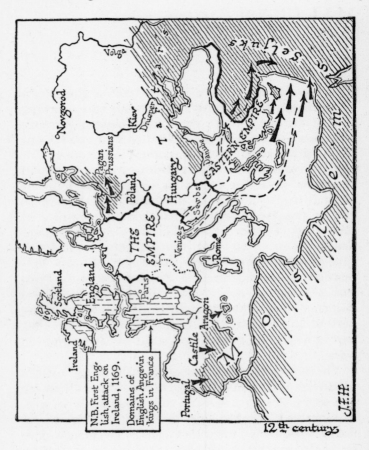

Volga

Novgorod

Kiev

Dnieper

Tartars

Pagan Prussians

Poland

Hungary

Danube

Serbs

EASTERN EMPIRE

Seljuks

THE EMPIRE

Venice

Rome

Scotland

England

Paris

Ireland

N.B. First English attack on Ireland, 1169.

Domains of English Angevin kings in France

Portugal

Castile

Aragon

M O O R S

12ᵗʰ century

J.F.H.

18

THE CHURCH
MILITANT

WE HAVE HAD no space here to tell of the rivalry
between Popes and Emperors for the real overlord-
ship of Europe. (The rulers of the eastern—German
—portion of Charlemagne's old dominions had again,
by the grace of the Pope, become Holy Roman Emperors. *Cf*. map 29.) The twelfth century saw a series
of events which bear witness to the power of the
Church in the Europe of that day. Into the eastern
end of the Mediterranean had come new invaders,
the Seljuk Turks, a barbarian people who had defeated the Arabs and established themselves in Palestine, Syria, and Asia Minor. All western Europe now
responded to a call from the Pope to rescue the Holy
Places from the infidel. Crusade after crusade was
launched, and the kings and princes (and large numbers of the peoples) of north-western Europe took
part in what were essentially a series of Mediterranean
wars.

There were crusades, too, not only in the eastern
Mediterranean. The Pope blessed the Christian
armies of Portugal, Castile and Aragon, who were
driving back the Moslems in Spain; and in the far
north he sanctified the wars waged by German knights
on the pagan Prussians of the Baltic shores, and by
Norman English nobles on the Christian, but non-
dues-paying, Irish.

MAP 10

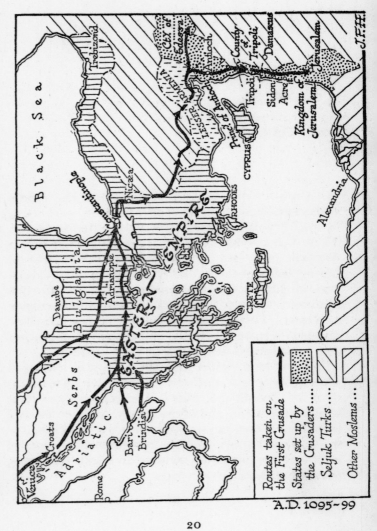

A.D. 1095~99

20

THE CRUSADERS

As a result of the Holy Wars against the infidel, various small states, none of them lasting, were set up by the Christian victors in the eastern Mediterranean. There was a kingdom of Jerusalem, a principality of Antioch, and counties of Tripoli and Edessa. But the lasting result of the Crusades lay not in these ephemeral kingdoms and principalities, but in the new ideas and outlook taken back by returning crusaders to western Europe; and in the great new merchant power of Venice which was enormously strengthened and consolidated as a result of the Crusades.

In fact, long before the last Crusades, crusaders were fighting one another even more wholeheartedly than they fought the infidel. And the result of their internecine strife—and of the steady, commercial realism of the Venetian republic—was the weakening of the Eastern Empire, and the consequent certainty of its crumbling before a new wave of Turkish invaders, the Ottomans, a century later.

MAP 11

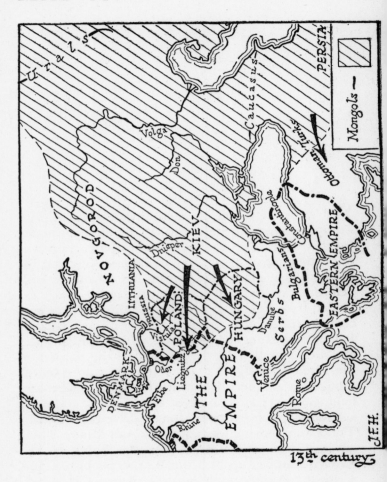

Mongols —

PERSIA

Urals

Volga

Don

Caucasus

Ottoman Turks

Dnieper

KIEV

NOVGOROD

LITHUANIA

Prussia

Vistula

Oder

Liegnitz

POLAND

Elbe

Rhine

DENMARK

THE

EMPIRE

HUNGARY

Danube

Serbs

Bulgarians

Constantinople

EASTERN EMPIRE

G.R.

Venice

Rome

J.F.H.

13th century

22

HORDES FROM ASIA

YET ANOTHER wave of alien invaders swept into Europe in the thirteenth century: the Mongols under Jenghis Khan. The original centre of Jenghis's power was thousands of miles away in eastern Asia; but his hordes of horsemen, well-armed and organised, had, before his death, won for him an empire which extended from Peking to Kiev on the Dnieper. His successor pressed still further into Europe, ravaged Poland and Hungary, and defeated an army of Poles and Germans at Liegnitz, just within the borders of the Empire itself, in 1241. Then the tide receded, though for another three centuries Mongol Khanates held sway over large areas of Russia.

Southward of the main Mongol drive the Ottoman Turks were moving westward into Asia Minor. Their advance was to continue—over the ruins of the Eastern Empire, up into the Balkan peninsula, and across the Danube itself into the plains of Hungary.

MAP 12

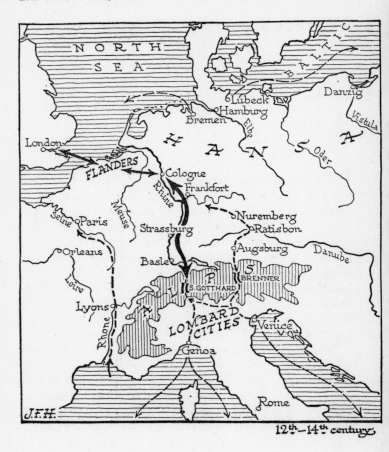

THE RISE OF THE TOWNS

We have already noted, in telling of the Crusades, the rise of a great merchant republic, Venice, whose wealth was based on trade and not upon land ownership. From the twelfth century onwards there had been growing up in Italy, Germany and France, great trading towns, in which a new merchant class challenged the dominance of the feudal lords. At the head of the Adriatic, where sea-routes met the land routes over the Alps leading to northern Europe, there grew up a number of cities which, leagued together, actually defied the Emperor himself.

Meantime, in the north, a new inland sea area was being economically unified by the merchants of the great League of the Hansa. This was a league of German cities based on the ports of Bremen, Hamburg, Lubeck and Danzig. Between these two groups was the connecting link of the Rhine, leading to Flanders and to London, the north-western "terminus."

Everywhere this increasingly wealthy and powerful merchant class was being courted by monarchs as a valuable ally against the feudal nobles.

MAP 13

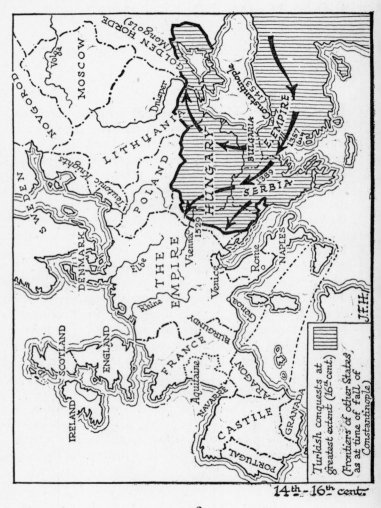

14th - 16th cent.

THE LAST INVASION
FROM ASIA

In the middle of the thirteenth century the Ottoman Turks, driven westward as we have seen by the
waves of Jenghis Khan's raiders, crossed the Dardanelles and entered Europe. Before the end of the
century they had broken the power of the Serbs,
crossed the Danube and occupied Wallachia, and
defeated the Bulgarians. Practically the whole of the
Balkan peninsula was in their possession before, in
1453, they took Constantinople and so ended the
Roman Empire of the East.

That event had far-reaching repercussions. In scat-
tering refugee scholars over western Europe the
Turks all unknowingly greatly assisted in the spread
of the New Learning; and their hold on the eastern
Mediterranean gave an added incentive to the Portuguese sailors who were pressing southward in the
search for a sea-route to the Indies by the west coast
of Africa.

In the sixteenth century the Turkish advance
northward was continued up to the very walls of
Vienna. The whole of Hungary, with the exception
of a small strip on the north-west, and nearly all the
Black Sea coast-line, were brought under the Sultan's
rule.

MAP 14

EMPIRE SHIFTS
WESTWARD

THE CRUSADES had turned men's thoughts to other lands and to visions of wealth overseas. The Portuguese voyages down the west coast of Africa had begun as expeditions against the infidel Moors. Less than half a century after the fall of Constantinople Vasco da Gama sailed round the Cape of Good Hope and opened the ocean way to India; and Columbus, heading due west with the same objective, had discovered a New World.

The opening-up of these new ocean routes made the Atlantic front of Europe for the first time the most important centre of development. Henceforth the Mediterranean and the Baltic were back streets. The countries fronting the open ocean now began to take a dominant place, not merely in European, but in world affairs. Lubeck and Hamburg, Venice and Genoa gave place to Amsterdam, Bristol, Lagos and Cadiz. At the outset, Spain and Portugal, the countries most closely in touch with the old Mediterranean world, took the lead. Spain, in particular, since her Habsburg rulers owned the Netherlands also, had the lion's share of the ports and trading towns which were henceforth to control the trade of the Seven Seas.

MAP 15

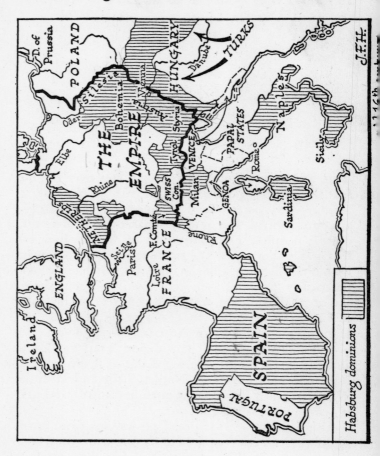

HABSBURG POWER

THE FINAL TRIUMPH of the monarchies over the old feudal nobility gave to certain ruling families in Europe—and particularly to one, the Habsburgs—absolute control over widely separate territories and diverse peoples. The Habsburgs of Austria had, ere this, made themselves the strongest of all the princes within the Empire, and the imperial title had become practically hereditary in that family. Their power was at its peak in the sixteenth century, in the reign of Charles V. Within the Empire he held the old Habsburg possessions of Austria, Styria, Tyrol, Bohemia and Silesia; and, in addition, Franche Comté and the Netherlands. He was King of Spain, and in Italy he ruled over Milan, Naples, Sicily and Sardinia. What remained of Hungary outside the area of Turkish conquest was his also.

Although these widespread possessions were never again after Charles' time held by one man, the Habsburg power in Europe remained for close on four more centuries—until the break-up of their empire by the Treaties of Trianon and St. Germain, 1919.

MAP 16

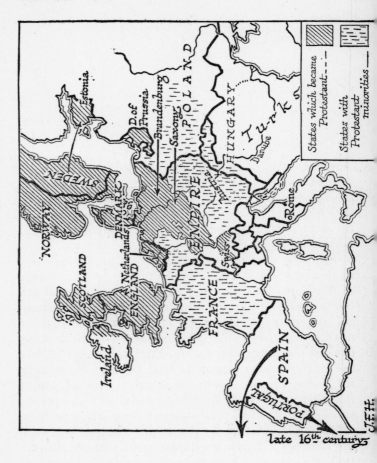

late 16th century.

J.F.H.

THE REFORMATION

THE GROWING rivalry between the old Mediterranean centres and the new national states of the north was accentuated by the great religious division of Europe which resulted from the Reformation. The Pope had, by decree, shared out the to-be-discovered world between the two Powers of Spain and Portugal. If the industrially developing north of Europe was to share in this new-found wealth it could only be by defiance of the Pope. Behind, or alongside, the quite real differences of opinion on religious belief and ecclesiastical organisation there lay the demand of these northern nations for a place in the sun. Before the close of the sixteenth century, England, Scotland, and the Scandinavian countries had become Protestant, while the Empire itself was roughly divided into a northern and central Protestant zone, with the possessions of the Catholic Habsburgs to east and west of it.

The Reformation finally ended the period when the Catholic Church gave some reality to the conception of a common European citizenship. It began the era of modern nationalism.

MAP 17

SCOTLAND

Cabot - Frobisher
to N. America

Ireland

ENGLAND

London

Plymouth

FRANCE

Flores

Azores

Vigo

PORTUGAL

SPAIN

Lagos

Cadiz

Madeira

Spanish route
to Central America

M o o r s

Canaries

Portuguese
route to India

Rio de Oro

S A H A R A

J.F.H.

late 16ᵗʰ century

34

THE BEGINNINGS OF
SEA POWER

With the opening-up of the new ocean routes a new factor appears in European—and world—history: sea power. The struggle between the northern nations and the Catholic Powers of Spain and Portugal inevitably led to a form of arbitrament previously unknown, at least on anything like so large a scale—fighting at sea. From the middle of the sixteenth century onwards, English privateers were infesting the routes of the Spanish treasure ships returning from Central America. Before the end of the century the defeat, by an English fleet composed of a vastly superior type of ship, of Philip of Spain's clumsy Armada, had definitely inaugurated the long chapter of England's dominance on the sea.

That dominance was challenged by the new state of the United Netherlands, itself the result of a revolt against Spain. Later still, England's chief rival at sea was France. But henceforth, though England might on occasion intervene directly in the affairs of the Continent, her main energies were always turned seawards.

MAP 18

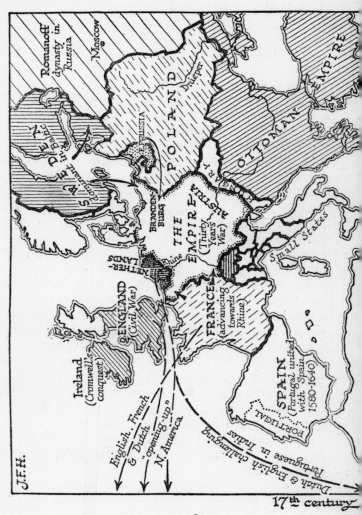

Romanoff dynasty in Russia

Moscow

RUSSIA

POLAND

Dnieper

SWEDEN

Baltic

OTTOMAN EMPIRE

BRANDEN-BURG

THE EMPIRE (Thirty Years' War)

AUSTRIA

HUNGARY

Rhine

NETHER-LANDS

Small States

ENGLAND (Civil War)

FRANCE (advancing towards Rhine)

Ireland (Cromwell's conquest)

SPAIN (Portugal united with Spain 1580-1640)

PORTUGAL

J.F.H.

English. French & Dutch "opening-up" N. America

Dutch & English challenging Portuguese in Indies

17th century

MODERN EUROPE
TAKES SHAPE

IF THE SIXTEENTH CENTURY saw the beginnings of the modern chapter of European history, the century which succeeded it saw the map of Europe taking on the broad dividing lines which we still know. The dominance of Spain was already waning. England, France and the Netherlands were establishing settlements and trading stations along the coast of North America, and challenging the power of the Portuguese in Asia. The former made certain that there should be no rival close to her western coast-line by carrying through, under Cromwell, a conquest of Ireland, which left that country drained and helpless.

But France has land as well as sea frontiers; and she was engaged in a struggle with the Empire which eventually brought the French eastern frontier nearer to the Rhine. The Empire itself was devastated by that most terrible of all the wars of religion, the Thirty Years' War; the end of which saw the independence of the Netherlands and of the Swiss Confederation formally recognised, and the emergence of Protestant Brandenburg in the north as a rival to Catholic Austria. Sweden was dominant on both shores of the Baltic. The Romanoff dynasty was established in Moscow. Italy still remained a mosaic of small states.

MAP 19

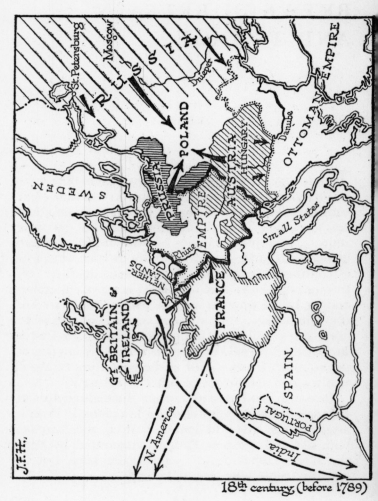

18ᵗʰ century (before 1789)

J.F.H.

38

THE STRUGGLE
BETWEEN BRITAIN
AND FRANCE

THE DOMINANT FACTOR in the making of European
history during the first three-quarters of the eighteenth century was the struggle, on land, sea, and overseas, between Britain and France. That struggle,
indeed, took place largely outside of Europe alto-
gether—in North America, where Britain won half
a continent from France, and lost another half
through the shortsighted policy of her own statesmen;
and in India, where a British trading company laid
the foundations of a vast empire. These two states
are already typical of the modern European Power,
in that their actual territory in Europe is important,
not so much in itself, but as the governing centre of
vastly larger territories in other continents.

In central and eastern Europe the eighteenth century saw Brandenburg, now the kingdom of Prussia,
asserting her strength against Austria; and, together
with her and the new Russia inaugurated by Peter
the Great, attacking the territorially large but in-
ternally feeble kingdom of Poland.

MAP 20

later 18ᵗʰ century

40

TWO REVOLUTIONS

We have just seen Britain and France as the two protagonists in a world-wide struggle for empire. Before the close of the eighteenth century, events in each were to affect powerfully not only Europe itself but those far distant areas of the world which were more and more being brought within the sphere of European influence.

The discovery of steam power and its application to industrial machinery, followed by a whole series of mechanical inventions, began in the England of the later eighteenth century that process—continued in our own day by the application of electrical power—to which we give the name of the Industrial Revolution. The results of that revolution, which spread (and is still spreading) over the whole world, were inevitably social and political, as well as economic.

In the last decade of the century a more obviously dramatic revolution took place in France; where an over-taxed peasantry and an oppressed city proletariat, led by a middle-class intent on establishing its power at the expense of the old nobility, beheaded a king and inaugurated a New Era, whose watchwords were to be Liberty, Equality, Fraternity. Modern democracy had been born; but it had still to reckon with the new lords of the world of industry.

N.B. British coalfields shown in the map include those (e.g. the Clyde and South Wales) which only came to be developed later.

MAP 21

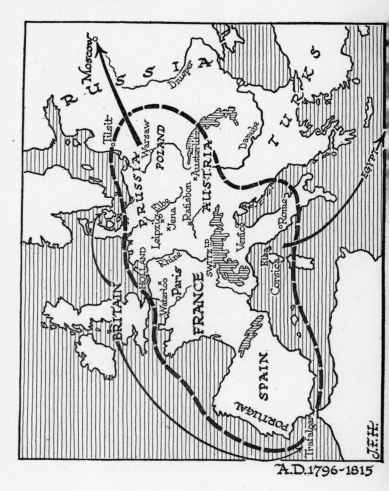

A.D. 1796~1815

NAPOLEON

Some men, by character or circumstance, write their names across a whole generation. Such a one was Napoleon Bonaparte, the Corsican lieutenant of artillery whom circumstance, *plus* an insatiable personal ambition, made the master of half Europe at the beginning of the nineteenth century. The early wars of the French Revolution were wars of defence against monarchical powers who wished to drown republicanism in blood. Under Napoleon, those wars became wars of conquest, although everywhere his armies marched with the watchwords of the new gospel of liberty and equality on their lips; and it is doubtful whether even his genius as a military commander could have won the victories associated with his name had not his troops regarded themselves as crusaders in a new cause.

Before the end of his career he had the greater part of the main peninsula of Europe under his control. Outside his sphere was Britain, whose sea power largely contributed to his downfall: and Russia, which was responsible, by her sheer size, for the most terrible of all his defeats.

43

MAP 22

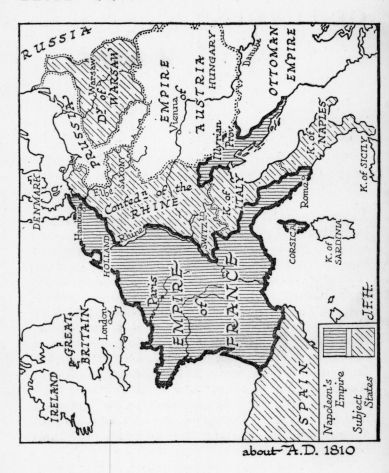

about A.D. 1810

44

"ROLL UP THE MAP OF EUROPE"

Under Louis XIV, France's eastern frontier had been pushed as far as the Upper Rhine. Napoleon carried it to the whole line of the Rhine and, in the north, beyond the Rhine, to the mouth of the Elbe. More than that, he broke up finally the thousand-year-old Holy Roman Empire. The little kingdoms, principalities, duchies, and cities of middle Germany were amalgamated into the Confederation of the Rhine. East of the Elbe was Prussia, shorn of the greater part of the territories which she had annexed from Poland; and, to the south, the Habsburg empire of Austria. Poland he resuscitated as the Duchy of Warsaw. The little states of Italy were in part absorbed into the empire of France, in part amalgamated into a new kingdom of Italy (the kingdom of Naples being still separate).

The old map of Europe had indeed, as Pitt feared, been "rolled up." But only temporarily.

MAP 23

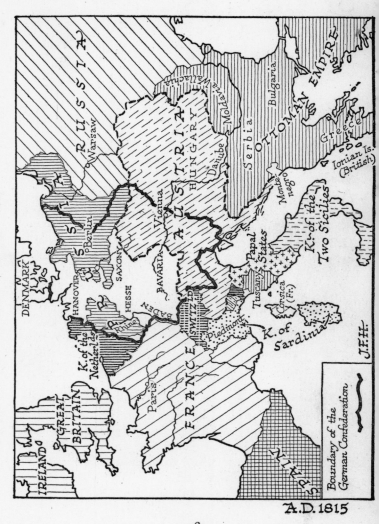

A.D. 1815

THE HOLY ALLIANCE

AFTER THE FINAL DEFEAT of Napoleon in 1815 the monarchs and statesmen of Europe hastened to redraw the map of Europe on the old dynastic lines. But the Holy Roman Empire had gone. Its place was taken by a German Confederation in which Prussia and Austria were rivals for the leadership. Each ruled areas outside the actual borders of the Confederation. Russia secured the greater part of what had been the Duchy of Warsaw. Belgium was added to Holland to form the Kingdom of the Netherlands. Italy reverted to the patchwork of small states which had existed before Napoleon; Lombardy and Venice going to Austria.

The Austrian statesman, Metternich, who redrew the map of Europe, looked upon the new Holy Alliance of emperors, kings and princes as the final safeguard of the re-established Old Order. A Concert of Europe in some measure foreshadowed the League of Nations of a century later.

MAP 24

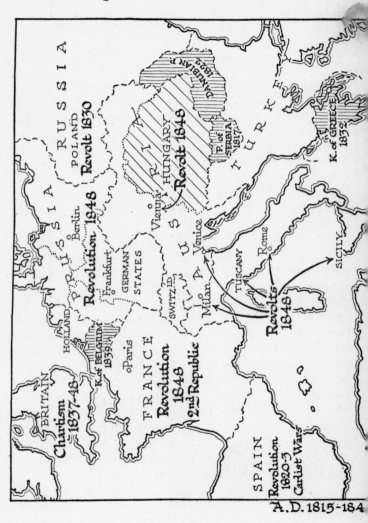

RUSSIA

PRUSSIA

POLAND Revolt 1830

Berlin

Frankfurt

GERMAN STATES

HOLLAND

K. of BELGIUM 1839

o Paris

FRANCE
Revolution
1848
2nd Republic

BRITAIN
Chartism
1837-48

SWITZ'D

AUSTRIA

HUNGARY
Revolt 1848

Vienna o

P. Venice

Milan

TUSCANY

Rome o

Revolts
1848

SICILY

DANUBIAN P. 1859

P. of SERBIA 1817

TURKEY

K. of GREECE 1832

SPAIN
Revolution
1820-3
Carlist Wars

THE SPECTRE OVER EUROPE

METTERNICH'S European system did not prove as stable as he had hoped. The years from 1815 onwards, culminating in the "year of revolutions," 1848, were marked by revolts against the *status quo* in every part of the continent. The new industrial proletariat rebelled against the tyranny of the machines; subject peoples everywhere, fired by the new nationalism and the doctrines of democracy which the French Revolution had broadcast over Europe, revolted against alien tyrants.

The Serbians rose against the Turks and gained a measure of autonomy in 1817; the Danubian principalities in 1822; the Greeks won their independence in 1832. The early 1820's saw revolution and civil war in Spain. In 1830 the Poles were in armed revolt against the Czar. In 1839 the Belgians broke away from the Netherlands and the Kingdom of Belgium was set up. Between 1837 and 1848 Chartism came near to becoming a revolutionary movement in Britain. 1848 saw revolutions in France, Germany, Hungary and Italy. In France the second republic was established, the Prince-President, Louis Napoleon, a little later making himself Emperor. Some of the German states, including Prussia, won short-lived constitutions. In Hungary and Italy the Habsburgs and their like maintained their positions.

MAP 25

HUNGARY

MOLDAVIA

RUSSIA

Odessa

SERBIA

WALLACHIA

Mouths
of the
Danube

CRIMEA

Sevastopol

BULGARIA

BLACK SEA

T

U

MACEDONIA

Constantinople

BOSPORUS

R

GREECE

DARDANELLES

K

Smyrna

E

Y

CRETE

CYPRUS

J.F.H.

Frontiers of 1854

mid-19th century

50

THE WESTERN POWERS
WATCH RUSSIA

INTERNATIONAL AFFAIRS in Europe in the middle
years of the nineteenth century were dominated by
the Eastern Question. Russia, which for little more
than a century had been a part of Europe proper,
was now a Power of terrifying size and resources,
with a frontier extending from the Baltic to the
mouths of the Danube. She was openly declaring her
intention of securing a sea outlet to the south by
ousting the Turk and establishing herself on the
Bosporus and the Dardanelles.

The Concert of Europe dreamed of a Balance of
Power which depended on no single state becoming
too powerful. The western Powers began to see in
the bolstering up of a much weakened Turkey the
best safeguard against Russian expansion. In the
Crimean War Britain and France were Turkey's
allies against Russia, and the Russian scheme of oc-
cupying the Danubian principalities was frustrated.
The age-old struggle between Christian and Moslem
had given place to one in which Catholic and Protestant Christians were allied with Moslems against
Greek Orthodox Christians.

MAP 26

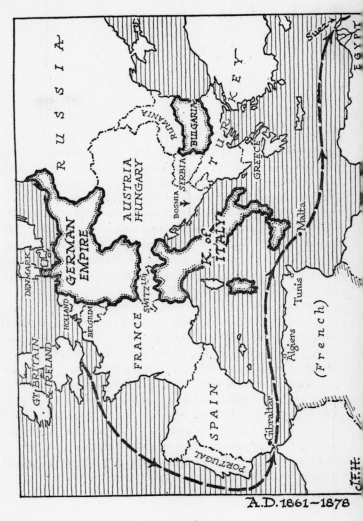

A.D. 1861-1878

J.F.H.

52

NEW POWERS IN
EUROPE

Between 1860 and 1880 the map of Europe was simplified by the inclusion of the various small Italian states in the kingdom of Italy (1861); and of the states of northern and central Germany under the leadership of Prussia in the German empire (1871). In 1878 another Russo-Turkish war led to the establishment of Bulgaria as an autonomous state under Russian protection, and to additions of territory, at the expense of Turkey, to Serbia and Greece. Russia's attempt to create a "Big Bulgaria," including most of Macedonia and Thrace, was checkmated by the Powers at the Congress of Berlin.

The year 1869 saw the Mediterranean once more made an area of first-class economic and political importance by the cutting of the Suez Canal. The route from Europe to India and the Far East now lay through the Mediterranean Sea, and Mediterranean politics inevitably became a matter of vital importance to Britain, whose most valuable imperial possessions lay in Asia. The new era of capitalist-imperialism was beginning; and the French annexation of Algeria and Tunis added to the strength of France's strategic position in the Mediterranean zone.

MAP 27

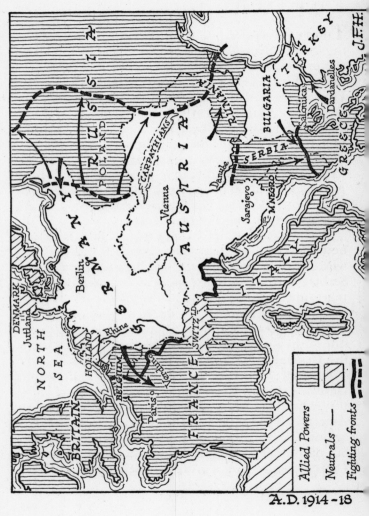

A.D. 1914-18

Allied Powers	Neutrals —	Fighting fronts

THE WORLD WAR

THIRTY YEARS of capitalist-imperialist rivalry between the great European Powers had made of Europe in the first years of the twentieth century a vast armed camp. There were two opposed groups. Britain and France, the erstwhile enemies of Russia, were now her allies. Against them was ranged the new Power of Germany, backed by Austria and Italy. German imperialism, baulked in its efforts to win as large a share of Asia and Africa as was held by the two western Powers, was alleged to be aiming at a "Mittel-Europa," a block of territory which was to extend from North Sea and Baltic, through the Balkans and Asia Minor, to the Persian Gulf.

War broke out in August 1914. The Habsburg heir-apparent had been murdered; it was declared, with Serbian complicity. Austria declared war on Serbia, Russia mobilised to protect her Balkan protégée, and within a day or two Germany, France, Belgium, and Britain were added to the combatants. Italy stood out, coming in a few months later on the side of the western allies. Bulgaria and Turkey joined the Central Powers.

For four years these latter, under the leadership of Germany, kept the war almost wholly on enemy territory. Only with the adhesion of the United States of America to the western allies did the forces arrayed against Germany become irresistible. Russia had gone out of the war in 1917, a revolution having ended the Czardom and set up a group of Socialist Republics.

MAP 28

NEW STATES AND
FRONTIERS

ONCE MORE, at the end of the Great War, the map of Europe was redrawn. Six new states appeared upon it —Finland, Estonia, Latvia, and Lithuania, all formed out of what had been Russian territory; Poland, consisting of pre-war Russian, German and Austrian possessions; and Czechoslovakia, which before the war had formed the northern part of the Habsburg empire.

Other large portions of that empire went to increase the areas of Rumania, Serbia (now Yugo-Slavia) and Italy. Austria itself was now a tiny inland state consisting of the earliest possessions of the first Habsburgs.

But the Habsburgs themselves had gone. Austria was a republic. The Romanoffs had gone from Russia, and the Brandenburg-Hohenzollerns from Germany. . . .

MAP 29

K. of DENMARK

FRISIA

D.y of SAXONY

Elbe

NORTH MARK

Brandenburg

Wends

(Prussians)

D.y of POLAND

Vistula

EAST MARK

THURINGIA

Rhine

D.y of LORRAINE

Oder

D.y of FRANCONIA

MEISSEN MARK

D.y of BOHEMIA

MORAVIA

Paris

K. of FRANCE

D.y of SWABIA

Danube

D.y of BAVARIA

Mk. of AUSTRIA

Vienna

K. of HUNGARY

D.y of CARINTHIA

K. of BURGUNDY

Rhone

LOMBARDY

Po

Venice

Drave

Danube

Save

K. of CROATIA

SERVIA

CORSICA

Moors

SARDINIA

PAPAL STATES

o Rome

ITALY

J.F.H.

0   250   500   750 Miles

10th century

58

GERMANY: AUSTRIA: (i) THE EMPIRE OF OTTO THE GREAT

THE EASTERN portion of Charlemagne's empire, the kingdom of the East Franks (*cf*. map 6), broke up into separate dukedoms during the ninth century. Then, early in the tenth century, a Duke of Saxony established his predominance over most of the German-speaking peoples. This ruler, Henry the Fowler, was succeeded in 936 by his son, Otto, who consolidated his father's work of unification; and, in return for assistance given to Pope John XII, was crowned by that prelate as Emperor of the Holy Roman Empire in 962.

The eastern frontier of the Empire was a more or less permanent battle line against the pagan Slav and Turanian peoples. Otto established a series of marks, or frontier governments, to defend the border and to press further eastward. Note especially the two which were the nuclei of the two states destined to play leading parts in the future history of the Empire—the North Mark (Brandenburg) on the Elbe; and the Mark of Austria, on the Danube.

MAP 30

HANSEATIC
LEAGUE
growing

K. of DENMARK

Teutonic
Knights

Lübeck
Hamburg

POMERANIA

Vistula

K. of
POLAND

Bremen
Munster

HOLLAND

BRANDEN
BURG

Elbe

Oder

SILESIA

BRABANT

Rhine

Cologne

Coblenz

Mainz

Nuremburg

BOHEMIA

MORAVIA

FRANCE

Strassburg

Ulm

Danube

Augsburg

AUSTRIA

Vienna

Basle

Everlasting
League

CARINTHIA

ALPS

Milan

Rhone

CARNIOLA

Venice

Drave

Danube

K. of HUNGARY

Save

PROVENCE

Genoa

PAPAL
STATES

SERVIA

Rome

Territory gained by
Rudolph of Habsburg,
1278.

By Teutonic Knights'
Crusade, c. 1230.

J.F.H.

13th century

GERMANY: AUSTRIA: (ii) TOWNS AND EMPERORS

THE GREAT DUCHIES which had formed the component parts of Otto the Great's empire began, in the twelfth century, to break up into a large number of smaller states (the frontiers of which are much too complicated to attempt to show in a single map). This process of disintegration—the very opposite of what was happening in the rest of north-western Europe at this time—was encouraged by the emperors in order to reduce the power of possible rivals. At the same time, they fostered the growth all over Germany of trading towns.

The main points to note in this period are: (1) the crusade carried on by the Teutonic Knights against the heathen Prussians of the Eastern Baltic; (2) the formation of the Hanseatic League (*cf.* map 59), beginning with the seaport towns of the northern coast and rapidly extending until it included most of the towns of the interior; (3) the beginnings of the modern state of Switzerland by the foundation (1291) of the Everlasting League of certain Alpine districts against the Habsburgs, their feudal lords (*cf.* map 46); and (4) the extension of the territories of these same Habsburgs by a war waged by Rudolph of Habsburg (1278) against the King of Bohemia, which established the Habsburgs as the rulers of Austria and the districts to the south of it.

MAP 31

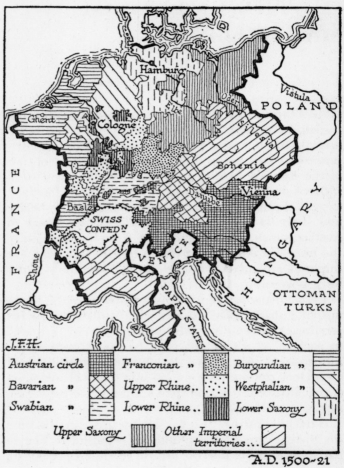

Austrian circle   Franconian "   Burgundian "
Bavarian   "   Upper Rhine ..   Westphalian "
Swabian   "   Lower Rhine ..   Lower Saxony

Upper Saxony   Other Imperial
                   territories ...

A.D. 1500~21

62

GERMANY: AUSTRIA:
(iii) THE EMPIRE OF
MAXIMILIAN I

THROUGHOUT the fourteenth and fifteenth centuries
the disintegration of the German lands proceeded
apace, although the nominal unity under the Emperor was still maintained, and the actual frontiers
of the Empire slightly extended both east and west.
Most of the great towns were linked by the Hanseatic
League. For the rest, the Empire was split into nearly
300 virtually independent states. In 1438 Albert of
Austria was chosen Emperor, and from that date, until
the final dissolution of the Empire by Napoleon, the
imperial crown was practically an hereditary Habsburg possession.

Early in the sixteenth century the Emperor Maximilian I made some attempt at a unification of his
territories by dividing the Empire into ten "circles,"
in each of which there would be some common ad-
ministration. An idea of the previous multiplicity of
German states can be gathered by a study of this map,
which illustrates Maximilian's attempt at consolida-
tion.

The Everlasting League had by this time become
the Swiss Confederation; and the republic of Venice
had gained a considerable area in territory, previously
part of the Empire, on the Italian mainland.

MAP 32

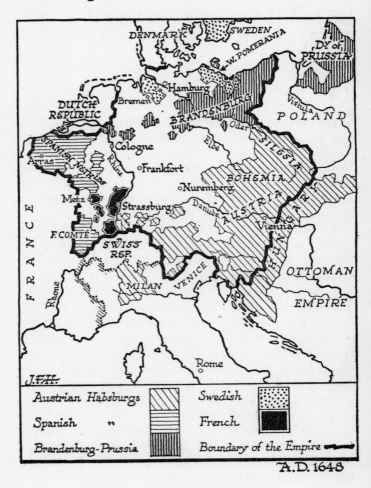

DENMARK

SWEDEN

Dy. of
PRUSSIA

W. POMERANIA

Hamburg

Bremen

BRANDENBURG

DUTCH
REPUBLIC

Vistula

POLAND

Oder

SILESIA

Cologne

Rhine

Spanish Netherlands

Arras

Frankfort

BOHEMIA

Nuremberg

AUSTRIA

Metz

Elbe

Strassburg

Danube

HUNGARY

F. COMTÉ

Vienna

FRANCE

SWISS
REP.

OTTOMAN

MILAN

VENICE

EMPIRE

Rhone

J.F.H.

Rome

Austrian Habsburgs   Swedish

Spanish   ,,   French

Brandenburg-Prussia   Boundary of the Empire

A.D. 1648

GERMANY: AUSTRIA: (iv) THE EMPIRE AT THE TREATY OF WESTPHALIA, 1648

Nowhere in Europe did the Reformation (*cf.* map 16) lead to such a bitter struggle as in the German lands. That struggle culminated in the Thirty Years' War, which began in 1618 with the refusal of the Protestant nobles of Bohemia to accept Ferdinand of Habsburg as their king, and went on until the population of Germany had been halved.

By the Treaty of Westphalia which ended it (1648) the power of the Empire was considerably reduced. The independence of the republics of Holland and Switzerland, already accomplished, was at last formally recognised. Considerable areas of the Empire were now held by outside states; Western Pomerania and the country to the west of the mouth of the Elbe, by Sweden; Alsace and certain bishoprics in Lorraine by France; and the Netherlands by the Spanish Habsburgs. The Imperial frontiers now ran north of the Alps, the Italian possessions of the Habsburgs being outside the Empire.

The one German prince who, in the course of the struggle, greatly increased his power and possessions was the Hohenzollern Elector of Brandenburg, a Lutheran. Henceforward, Brandenburg-Prussia was the most formidable German Power next to Austria.

MAP 33

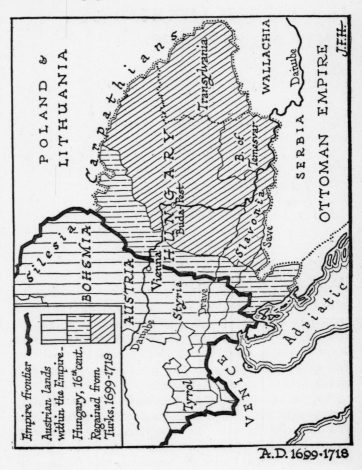

POLAND & LITHUANIA

Carpathians

WALLACHIA

Danube

SERBIA

OTTOMAN EMPIRE

J.F.H.

Transylvania

B. of Temesvar

Silesia

BOHEMIA

AUSTRIA

HUNGARY

Buda-Pest

Slavonia

Save

Vienna

Styria

Drave

Danube

Tyrol

VENICE

Adriatic

Empire frontier

Austrian lands within the Empire.

Hungary, 16th cent.

Regained from Turks, 1699-1718

A.D. 1699-1718

GERMANY: AUSTRIA: (v) THE HABSBURGS AND HUNGARY

W<small>HILE</small> Brandenburg-Prussia was consolidating its power in northern Germany, the Habsburgs were greatly extending their territories outside the boundaries of the Empire. In 1556 Austria, Hungary, and Bohemia were united with the Imperial crown. At that date by far the larger part of Hungary was in the hands of the Turks (*cf.* map 68).

A series of campaigns in the later seventeenth century pushed the Turks back southwards towards the Save and the Danube. By 1699 Hungary proper and Transylvania had been reconquered, and by 1718 Slavonia and the Banat of Temesvar were added to the Habsburg dominions.

MAP 34

North Sea

SWEDEN

DENMARK

Baltic

Elbe

DUTCH REPUBLIC

HANOVER

Berlin

POLAND

Vistula

Warsaw

SILESIA

SAXONY

FRANCE

Strassburg

Rhine

BAVARIA

Danube

Vienna

HUNGARY

SWISS REP.

Drave

Rhone

Save

OTTOMAN EMPIRE

Rome

J.F.H.

Austrian   Prussian   Boundary of the Empire

A.D. 1789

GERMANY: AUSTRIA: (vi) THE RISE OF PRUSSIA

THE EIGHTEENTH CENTURY saw the power and influence of Prussia among the German peoples enormously increased. By a series of wars Frederick II of Prussia (called the Great), who reigned from 1740 to 1786, took Silesia from Austria; and by sharing with Russia and Austria in the first partition of Poland (1772) he acquired the coastlands of the Lower Vistula, thus linking his Brandenburg possessions with Prussia proper (*cf.* map 32). Prussian dominance in North Germany was also consolidated by the acquisition of various territories between the Rhine and Elbe, and on the Rhine itself.

On the Empire's western frontier considerable territory was lost in the wars with Louis XIV of France; French Flanders was regained by France, and by taking Lorraine, Alsace and Franche Comté Louis pushed the French frontier right up to the Rhine. Meantime, by the Treaty of Utrecht (1713) the Netherlands passed from Spain to Austria.

MAP 35

North Sea

DENMARK

Baltic

SCHLESWIG

HOLLAND

Elbe

HANOVER

Berlin

Vistula

RUSSIA
(Poland)

BELGIUM

Cologne

SAXONY

Oder

LORRAINE

ALSACE

Rhine

BADEN

WÜRTTEM-
BERG

BAVARIA

Danube

Vienna

FRANCE

Drave

HUNGARY

SWITZ.LD

VENETIA

Adriatic

Rhone

TURKS

J.F.H.

Austrian Empire

Km. of Prussia

Boundary of the Confederation
of 1815

"   of the North German
Confed.n, 1866

A.D. 1815-1866

GERMANY: AUSTRIA:
(vii) MODERN GERMANY
TAKING SHAPE

THE RAMSHACKLE Holy Roman Empire was finally
ended by the conquest of Napoleon. It was formally
dissolved in 1806. In that same year Austria was
definitely separated from the rest of Germany, although in 1815, after the final defeat of Napoleon,
when 39 sovereign German states formed themselves
into a Confederation, the German part of the Austrian Empire was again included (*cf.* map 23).

Prussia, reconstituted after her crushing defeats by
Napoleon, was once again dominant in the north.
She now held most of the Rhineland and by the
acquisition of Northern Saxony had extended her
territories on the Elbe. By the formation of a Zoll-
verein (Customs Union) in 1834 she succeeded in
extending her influence over all the northern states.
In 1864, under Bismarck's leadership, she took Schleswig from Denmark. Two years later she declared the
Confederation dissolved, and after a short, sharp war
with Austria asserted once and for all her supremacy
in Germany. She formed (1866) the North German
Confederation, annexing Hanover, Hesse-Nassau and
other territories, and thus unifying her two blocks
of possessions. The only states left out of the new
Confederation were Baden, Württemberg and Bavaria, all three in the south.

MAP 36

DENMARK

SCHLESWIG

HOLLAND

BELGIUM

ALSACE-
LORRAINE

SWITZ.ᴸᴰ

P R U S S I A

Elbe

Berlin

Oder

Vistula

Rhine

BADEN

WÜRTTEM-
BERG

BAVARIA

Vienna

Danube

J.F.H.

Boundary of the
German Empire,
1871

The Holy Roman Empire
at its greatest extent
(north of the Alps) ....

A.D. 1871

GERMANY: AUSTRIA: (viii) THE NEW GERMAN EMPIRE

Bɪsᴍᴀʀᴄᴋ's plans for the unification of the new Germany were completed by the war with France (1870). In the Hall of Mirrors at Versailles the King of Prussia, in the presence of the other German monarchs, was proclaimed German Emperor. The three southern states of Baden, Württemberg and Bavaria formed part of the new empire, and the imperial crown was made hereditary in the Hohenzollern monarchy of Prussia. The provinces of Alsace and Lorraine were taken from France, and added as a separate state to the imperial federation.

The map shows the extent of this new German empire as compared with the territories of the Holy Roman Empire north of the Alps.

MAP 37

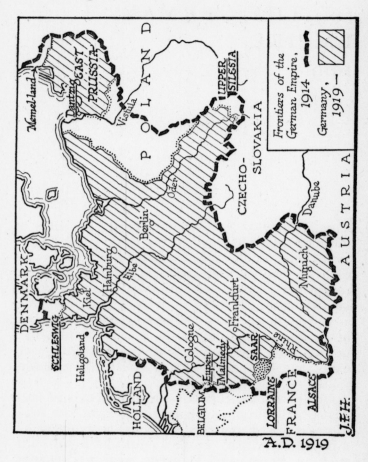

A.D. 1919

GERMANY: AUSTRIA: (ix) THE GERMAN EMPIRE, 1919

THE TREATY OF VERSAILLES destroyed the unity of the territories combined in the German Empire of 1871. Once again, as before Frederick the Great's time (*cf.* map 32), East Prussia was isolated from the rest of the Prussian territories by a strip of Polish land along the line of the Vistula. The frontier line, as will be noted, to-day follows hereabouts the line of the 17th and 18th centuries: except that in 1919 a portion of Upper Silesia was detached from Germany, and divided between Poland and Czechoslovakia.

On her western frontier Germany had to hand back Alsace and Lorraine to France, while the Saar territory was placed under international government for a period of 15 years. Further north the frontier districts of Eupen and Malmèdy went to Belgium. Northern Schleswig went back to Denmark.

75

MAP 38

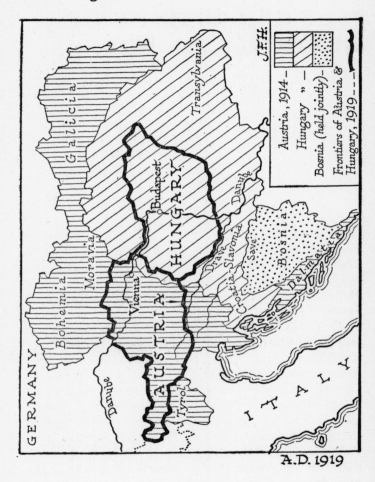

GERMANY

Galicia

Bohemia

Moravia

Vienna

AUSTRIA

Danube

Tyrol

Transylvania

Budapest

HUNGARY

Drave

Danube

Croatia-Slavonia

Save

Bosnia

Dalmatia

ITALY

J.F.H.

Austria, 1914 —
Hungary " —
Bosnia (held jointly) —
Frontiers of Austria &
Hungary, 1919 — — —

A.D. 1919

GERMANY: AUSTRIA: (x) THE BREAK-UP OF AUSTRIA-HUNGARY, 1919

THE HABSBURG empire had been still further enlarged by the acquisition of the Polish territory of Galicia at the end of the eighteenth century; and again by the occupation of the Turkish province of Bosnia-Herzegovina in 1878 (formally annexed in 1908).

This elaborate structure of widely varying peoples and territories was ruthlessly broken up by the Peace Treaties of 1919 (the Treaty of St. Germain with Austria, and the Treaty of Trianon with Hungary). Out of the Austro-Hungarian lands the new state of Czechoslovakia, in the north, was constituted; the reformed state of Poland took Galicia; while, on the south, large areas went to Rumania, Serbia (now Yugo-Slavia) and Italy. To Austria was left little more than the old Archduchy of Austria lying within the Holy Roman Empire (*cf.* map 30).

MAP 39

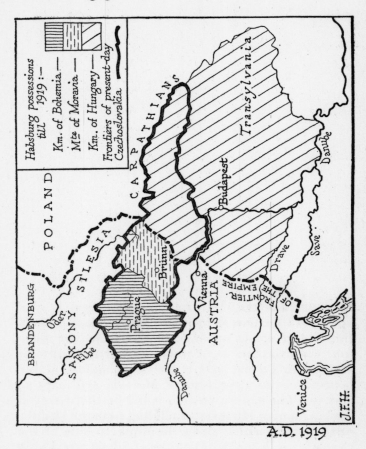

Habsburg possessions
till 1919 :—
Km. of Bohemia.
Mts of Moravia.
Km. of Hungary.
Frontiers of present-day
Czechoslovakia.

POLAND

BRANDENBURG

SAXONY SILESIA

Oder

Elbe

CARPATHIANS

Transylvania

Brünn

Prague

Danube

Budapest

Danube

Vienna

AUSTRIA

Drave

FRONTIER OF THE EMPIRE

Save

Danube

Venice

J.F.H.

A.D. 1919

CZECHOSLOVAKIA, 1919

By the treaties which followed the Great War of 1914–1918 the new state of Czechoslovakia was established. It is made up of territories which for some four centuries had been possessions of the house of Habsburg. To the west is the old kingdom of Bohemia and the margravate of Moravia (*cf.* maps 29 and 30). The eastern portion is the northern area of the kingdom of Hungary, extending from the Danube just east of Vienna, along the southern side of the Carpathians. A small area of Upper Silesia, previously German, was also placed within the Czechoslovak frontiers.

MAP 40

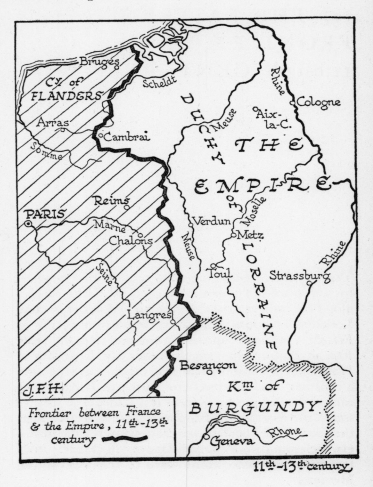

Bruges

Cy of
FLANDERS

Scheldt

DUCHY

Rhine

Cologne

Meuse

Aix-
la-C.

Arras

Cambrai

THE

Somme

EMPIRE

Reims

Meuse

PARIS

Verdun

Moselle

Marne

Metz

Chalons

Meuse

Seine

Toul

Strassburg

Rhine

LORRAINE

Langres

J.F.H.

Besançon

Km of
BURGUNDY

Frontier between France
& the Empire, 11th–13th
century

Geneva

Rhone

FRANCE:
(i) THE EASTERN FRONTIER, 11th-13th CENTURY

Wʜᴇɴ Cʜᴀʀʟᴇᴍᴀɢɴᴇ's empire was divided between his three sons, the central area west of the Rhine, along the valleys of the Meuse and Moselle, became Lothar's kingdom (in German, Lothringen; in French, Lorraine). And as, on the one side, the Holy Roman Empire, and, on the other, the Kingdom of France, were consolidated, this middle area became a constant battleground. The Duchy of Lorraine became a part of the Empire, as did ultimately, further south, the Kingdom of Burgundy. In the Middle Ages the County of Flanders, covering what is now northern Belgium, was a part of France, the French frontier running thence southwards to the west of the Meuse.

MAP 41

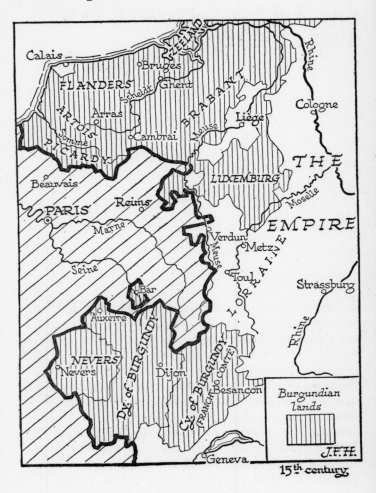

Calais · FLANDERS · Bruges · Ghent · ZEELAND · BRABANT · Rhine · Cologne · Scheldt · ARTOIS · Arras · Cambrai · Meuse · Liège · THE · PICARDY · Somme · LUXEMBURG · Beauvais · Reims · Moselle · EMPIRE · PARIS · Marne · Verdun · Metz · Seine · Meuse · Toul · LORRAINE · Strassburg · Bar · Rhine · Auxerre · NEVERS · Dy. of BURGUNDY · Dijon · Ct. of BURGUNDY (FRANCHE COMTÉ) · Besançon · Nevers · Geneva

Burgundian lands

J.F.H.

15th century

82

FRANCE:
(ii) THE BURGUNDIAN POSSESSIONS

IN THE fifteenth century it seemed highly probable that Burgundy would be permanently established as a separate state between the Empire to the east, and France to the west. Charles the Bold (1467–77) ruled over not only the territories of Burgundy proper, to the west of what is now Switzerland, but over Flanders, Brabant and Luxemburg, and the French provinces of Artois and Picardy. Before the end of the century, however, the Duchy of Burgundy, with Picardy and Artois, went back to France; while, through the marriage of Charles' daughter to Maximilian, the Netherlands reverted to the Empire.

MAP 42

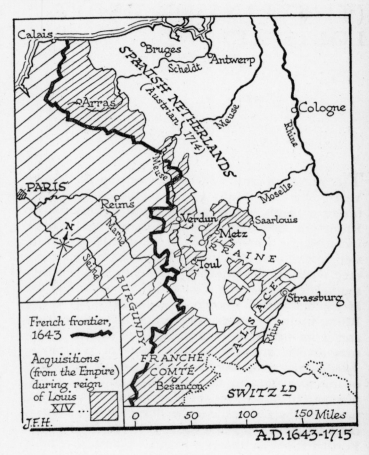

Calais

Bruges

Antwerp

Scheldt

SPANISH NETHERLANDS
(Austrian 1714)

Arras

Meuse

Cologne

Rhine

Meuse

PARIS

Reims

Marne

Verdun

Saarlouis

Moselle

Metz

LORRAINE

Seine

N

Toul

BURGUNDY

Strassburg

ALSACE

Rhine

FRANCHE
COMTÉ
Besançon

SWITZ^LD

French frontier,
1643 ——

Acquisitions
(from the Empire)
during reign
of Louis
XIV ...

J.F.H.

0   50   100   150 Miles

A.D. 1643-1715

FRANCE:
(iii) EXPANSION UNDER
LOUIS XIV

THE DISINTEGRATION of Germany during the Thirty
Years' War (1618–48) gave to France the opportunity to push her frontier eastwards towards the
Rhine. In the north the (Habsburg) Netherlands
included the whole of Flanders, running to a line
well to the west of Arras. A large part of this territory was now reconquered by the French. Further
south, the frontier was advanced from the line of the
Meuse across the Moselle, and in Alsace up to the
Rhine itself. The three bishoprics of Verdun, Metz
and Toul had been occupied by the French a century earlier, in 1552. By the Treaty of Westphalia
(1648) they went to France, though still nominally
owing suzerainty to the Holy Roman Empire (*cf.*
map 32). In Alsace also, the landgraviate of which was
now won by the French, the Emperor still claimed
certain rights. Strassburg itself was occupied in 1681.
Further to the south again Franche Comté, originally
a part of the kingdom of Burgundy, was added to the
French domains.

MAP 43

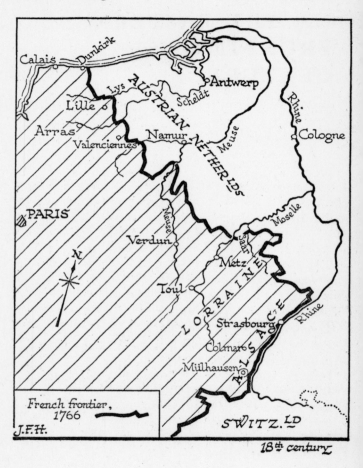

Calais Dunkirk
Lille
Arras
Valenciennes
Lys AUSTRIAN
Scheldt •Antwerp
Namur NETHER·L⁴ˢ
Meuse
Rhine
•Cologne

PARIS

Verdun
Meuse
Toul
Metz
Saar
Moselle
L O R R A I N E
Strasbourg
Colmar
Mülhausen
A L S A C E
Rhine

N

French frontier,
1766

J.F.H.

SWITZ.ᴸᴰ

18ᵗʰ century

86

FRANCE:
(iv) THE EASTERN FRONTIER, 18th CENTURY

BEFORE THE END of the eighteenth century the eastern frontier of France had been advanced to, and consolidated upon, a line running south-eastwards from the neighbourhood of Dunkirk to a point on the Rhine north of Strasbourg. Lorraine (i.e. the territory remaining outside the borders of Verdun, Metz and Toul) was occupied early in the century and finally annexed in 1766. The tiny independent republic of Mülhausen in Upper Alsace, which had been joined to the Swiss Confederation, was united to France by its own request in 1798.

MAP 44

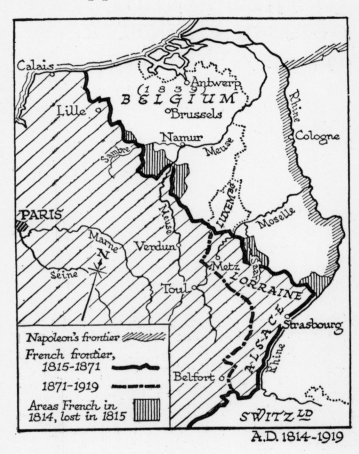

Calais

Lille

(1839)
BELGIUM
Antwerp
Brussels

Namur

Sambre

Meuse

Rhine

Cologne

PARIS

Meuse

LUXEMB^G

Marne

N

Verdun

Moselle

Seine

Metz

Toul

Saar

LORRAINE

Belfort

ALSACE

Rhine

Strasbourg

Napoleon's frontier
French frontier,
1815-1871
1871-1919
Areas French in
1814, lost in 1815

SWITZ^{LD}

A.D. 1814-1919

FRANCE:
(v) THE FRANCO-GERMAN FRONTIER, 19th-20th CENTURY

Napoleon advanced his frontier right up to the Rhine, from Switzerland to a point some distance north of Cologne. But on his defeat it was once again fixed on a line approximately the same as that of the eighteenth century. (Some slight additions to this were made in the original settlement of 1814—see map —but these were taken back from France in the following year.) The northern part of the frontier now of course divided France from the newly constituted kingdom of the Netherlands (Belgium, 1839).

After the war of 1870–71 Alsace-Lorraine was taken from France and made a possession of the new German Empire. In 1919 the provinces went back to France.

MAP 45

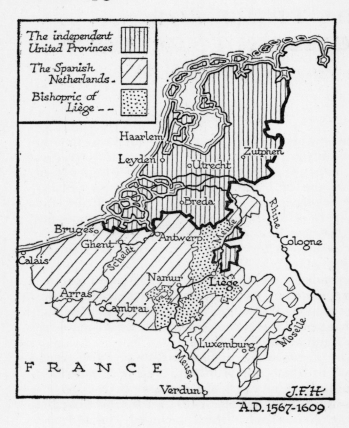

The independent United Provinces

The Spanish Netherlands.

Bishopric of Liège

Haarlem

Leyden

Zutphen

Utrecht

Breda

Bruges

Ghent

Antwerp

Scheldt

Meuse

Rhine

Cologne

Calais

Namur

Liège

Arras

Cambrai

Moselle

Luxemburg

F R A N C E

Meuse

Verdun

J.F.H.

A.D. 1567-1609

THE NETHERLANDS:
THE FIGHT FOR
INDEPENDENCE

THE LOW-LYING LANDS north and south of the mouths
of the Rhine and the Meuse had been for centuries
(*cf.* maps 40–44) possessions of France, Burgundy, or
the House of Habsburg. In the sixteenth century
they belonged to the Habsburg rulers of Spain; and
Philip II of Spain set himself the task of stamping
out the new Reformed religion in these territories.
The result was revolt. Forty years of bitter fighting,
during which (by the Union of Utrecht, 1579) the
independence of the northern provinces was declared, ended in 1609, when a twelve years' truce was
agreed upon. When war again broke out at the end
of that period the Netherlanders extended their territory south of the Rhine.

The Treaty of Westphalia (1648) gave full recognition to Dutch independence. Napoleon ended the
republic and incorporated Holland in the French
Empire. The settlement of 1815 set up the Kingdom
of the Netherlands, which included Holland and
Belgium (i.e. the Habsburg Netherlands and the
Bishopric of Liège). In 1830 Belgium revolted, and in
1839 a treaty of separation between Holland and
Belgium was signed.

MAP 46

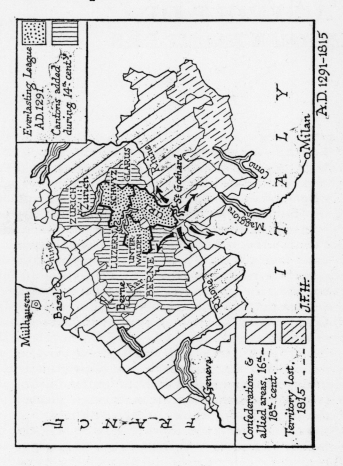

Everlasting League A.D. 1291

Cantons added during 14th cent.

A.D. 1291–1815

I T A L Y

o Milan

Como

Maggiore

Rhine

St Gothard

SCHWYZ

GLARUS

ZURICH

Zurich

LUZERN

UNTER WALDEN

BERNE

URI

Berne

Basel

Rhine

Mülhausen

Rhone

Geneva

F R A N C E

J.F.H.

Confederation & allied areas, 16th–18th cent.

Territory lost, 1815

92

SWITZERLAND:
A REVOLT OF
MOUNTAINEERS

In 1291 (*cf.* map 30) three small districts in the
High Alps revolted against their Habsburg over-
lord, and formed an "Everlasting League." Their
geographical position north of the key pass of St.
Gothard gave them control of important trade routes
between the Empire and the cities of the Lombard
Plain. During the following century other adjoining
areas joined the League, and an Austrian invasion
of their territory was defeated. Later wars with Austria and with Burgundy added yet further territories,
and by the sixteenth century the area of the Swiss
Confederation was approximately the same as that
of present-day Switzerland. The Treaty of Westphalia (1648) gave full European recognition of
Swiss independence; and the post-Napoleonic settlements, which took away from Switzerland a strip of
territory on the south-eastern frontier, guaranteed
the permanent neutrality of the country.

There is no Swiss "race." The inhabitants of
Switzerland speak French, German or Italian, ac-
cording to their geographical location on the north-
western, north-eastern, or southern slopes of the Alps.

MAP 47

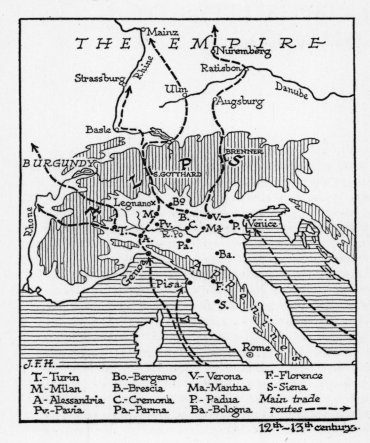

THE EMPIRE

Mainz

Nuremberg

Ratisbon

Strassburg

Rhine

Ulm

Danube

Augsburg

Basle

BURGUNDY

BRENNER

S. GOTTHARD

Rhone

Legnano

Bo

M

B.

V.

Pv.

C.

Ma

P.

Venice

T.

R. Po

A.

Pa.

Ba.

Genoa

Pisa

F.

S.

Rome

J.F.H.

T.- Turin	Bo.- Bergamo	V- Verona	F-Florence
M- Milan	B.- Brescia	Ma- Mantua	S- Siena
A- Alessandria	C.- Cremona	P.- Padua	*Main trade*
Pv.-Pavia	Pa.- Parma	Ba.- Bologna	*routes* ---→

12th – 13th century.

94

ITALY:
(i) THE CITIES OF
LOMBARDY

From the time of the break-up of the Roman Empire in the west until the nineteenth century, Italy was a geographical, not a political, expression. North and south of the territorial possessions of the Popes, which lay across the centre of the peninsula, Italy was cut up into various small states, often ruled by foreign dynasties.

But in the twelfth and thirteenth centuries one area of Italy was in the very forefront of the world scene. The cities of the Lombard Plain, lying between the seaports of Venice, Genoa and Pisa, and the trade routes running north and north-westward across the Alps to the towns of the Empire, struck the first blow against feudalism. Banded together in the Lombard League they defied the suzerainty of the Emperor and actually defeated him at the battle of Legnano (1176). Later, most of them were ruled by their own merchant princes elevated to the rank of sovereigns. But the great republics of Venice and Genoa remained for some centuries European Powers of considerable importance.

MAP 48

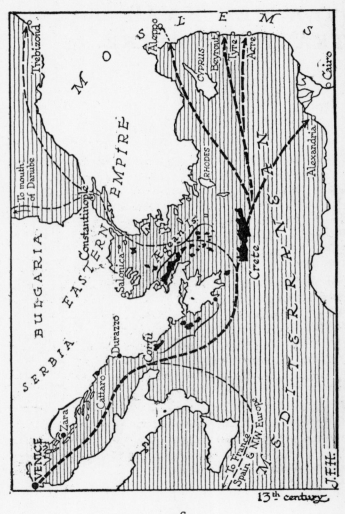

13th century

96

ITALY:
(ii) A MEDIÆVAL TRADING EMPIRE

VENICE, originally founded among the salt lagoons at the head of the Adriatic by refugees fleeing from northern barbarian invaders in the fifth century, steadily developed as a sea trading power during the centuries which followed. Cut off for long from any expansion on the mainland her burghers looked seaward. The spread of the Arab power along eastern and southern Mediterranean gave them the opportunity to act as intermediaries between Europe and the Moslem world. By the time of the Crusades she was strong enough to demand her full share of the spoils of conquest. Indeed, after the Fourth Crusade, during which the Eastern Empire and Constantinople itself were attacked, it was to Venice that the only lasting fruits of victory fell.

Her trade routes at this period covered all the eastern Mediterranean. She had ports or special trading facilities on every coast and many of the islands were in her possession.

MAP 49

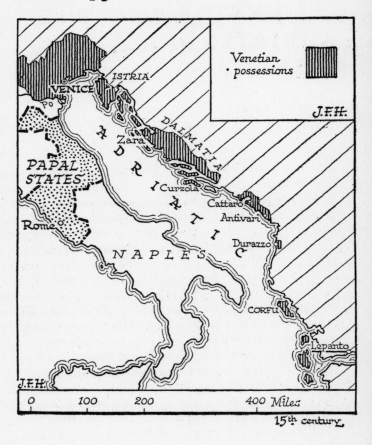

Venetian · possessions

J.F.H.

ISTRIA

VENICE

Po

DALMATIA

A D R I A T I C

Zara

PAPAL STATES

Curzola

Rome

Cattaro

Antivari

N A P L E S

Durazzo

CORFU

Lepanto

J.F.H.

0   100   200   400 Miles

15ᵗʰ century

98

ITALY:
(iii) VENICE AND THE ADRIATIC

By THE BEGINNING of the fifteenth century Venice
had reached the zenith of her power. She had expanded her possessions on the mainland and held the
greater part of the eastern Adriatic coast-line. But
before the end of the century her decline had begun.
The Turks had taken, or were threatening, her eastern possessions; Emperor and Pope alike were jealous
of the power of the republic and angered by the
bargains which her traders attempted to make with
the infidel; and above all, in the last decade of the
century, came the discoveries of the ocean routes to
the Indies and the New World which effectually destroyed the trading monopoly of the Venetians.

The republic lingered on until it was finally de-
stroyed—along with so many obsolete survivals—by
Napoleon. But for three centuries before that the
Adriatic had been a backwater, and the trade of the
Mediterranean relatively insignificant. The memory
of Venetian dominance on the eastern Adriatic coast
is a potent cause of jealousy between Italians and
Yugo-Slavians to-day.

MAP 50

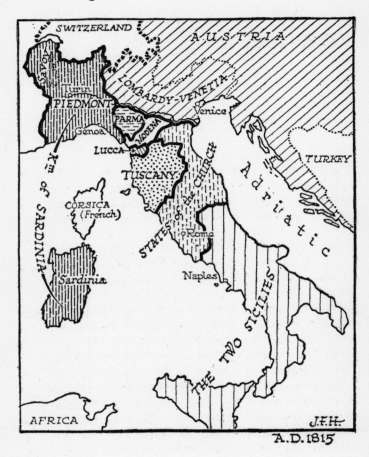

SWITZERLAND

AUSTRIA

LOMBARDY-VENETIA

Turin

PIEDMONT

Genoa

PARMA

MODENA

Venice

Lucca

TUSCANY

STATES of the CHURCH

CORSICA (French)

Adriatic

TURKEY

K⁰ of SARDINIA

Sardinia

Rome

Naples

THE TWO SICILIES

AFRICA

J.F.H.

A.D. 1815

ITALY:
(iv) THE BEGINNING OF
THE 19th CENTURY

THE CONGRESS OF VIENNA which re-drew the map of
Europe after the defeat of Napoleon (1815) con-
firmed anew the division of Italy into various small
states. The provinces of Lombardy and Venetia in
the north went to Austria, which exercised, indeed, a
dominating influence in the whole of the peninsula.
The possessions of the Pope still lay astride the centre
of the peninsula. There were, besides, the kingdom
of Sardinia, which included the island of Sardinia
and the provinces of Savoy and Piedmont; the kingdom of the two Sicilies in the south; and the duchies
of Modena, Parma, Tuscany, and Lucca.

MAP 51

SWITZ.LD · AUSTRIA

SAVOY

TRENTINO

LOMBARDY

VENETIA

PIEDMONT

Venice · Trieste

JUGO-

Fiume

PARMA

SLAVIA

Nice

Genoa

MODENA

LUCCA

Zara

TUSCANY · STATES of
the CHURCH

CORSICA

Rome

Naples

Sardinia

THE TWO SICILIES

K. of Sardinia, 1859
Annexed to Sardinia 1860
From Austria, 1866
Rome, 1870
Gained from Austria, 1919
Ceded to France, 1860

J.F.H.

A.D. 1859–1919

ITALY:
(v) THE UNIFICATION

THE NINETEENTH CENTURY saw the establishment at
long last of unified political control over the whole
of Italy. The revolutionary wave of 1848 gave a tre-
mendous impetus to the movement led by Garibaldi
and Mazzini for Italian independence. In 1859 Sardinia, in alliance with the French, defeated Austria
and the latter had to cede the province of Lombardy.
In the same year Modena, Parma, Lucca, Tuscany,
the two Sicilies, and a large part of the States of the
Church were annexed by Sardinia. Only the province
of Venetia and the territory of Rome now remained
outside the new kingdom. In 1866 Italy, allied with
Prussia in the war against Austria, recovered Venetia;
and in 1870, when a French garrison, sent by Louis
Napoleon to protect the Pope, was withdrawn, Italian
troops took possession of the city and the surrounding province.

Savoy and Nice had been ceded to France in 1860,
but after the Great War further territory was gained
by Italy from Austria in the Trentino and around
Trieste.

MAP 52

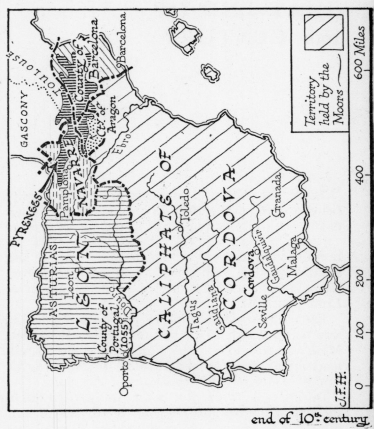

SPAIN:
(i) AT END OF
10th CENTURY

"SPAIN," OF COURSE, did not exist as a political entity before the end of the fifteenth century. The conquest of the peninsula by the Arabs in the eighth century (*cf.* map 5) was practically complete. Only in the far north-western corner, in Galicia and Asturias, did small Christian "kingdoms" maintain themselves. These were eventually unified in the kingdom of Leon (A.D. 866).

Eastward, along the line of the Pyrenees, in the area which Charlemagne had re-conquered from the Moslems and which under him became the Spanish mark, there grew up the kingdom of Navarre, the county of Aragon, and the county of Barcelona.

During these first centuries the Arab civilisation in the rest of the peninsula was far in advance of anything else in western and north-western Europe, and Christian students from France and Italy were present in considerable numbers in the University of Cordova.

MAP 53

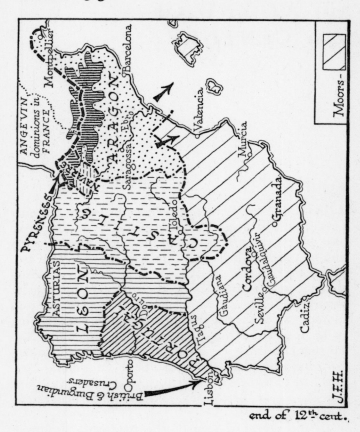

SPAIN:
(ii) AT END OF
12th CENTURY

D URING the twelfth century Crusaders were not only
fighting the Saracens in Palestine and Syria, but were
also, by express permission of the Pope, lending their
aid in pushing back the Arabs in Spain. The two
kingdoms of Leon and Castile were steadily pushing
their frontiers southward. In 1055–64 the district
around Oporto was won back from the Moors and
in 1095 the county of Portugal was given by the King
of Leon to Henry of Burgundy, a Crusader. In 1140
Portugal became a kingdom. Navarre remained,
covering the country round about Pamplona. Aragon had extended her territories down the valley of
the Ebro towards the sea, and had absorbed Barcelona.

In the southern part of the peninsula the great
caliphate of Cordova, attacked by various new Moor-
ish invasions from Africa, had given place to a
number of smaller emirates.

MAP 54

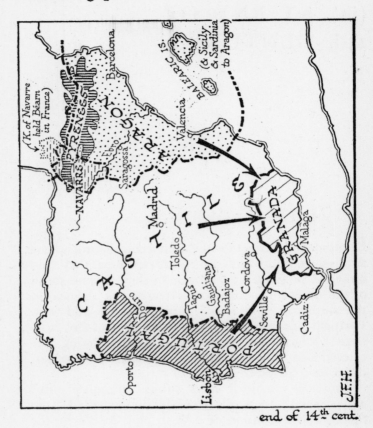

(K. of Navarre
held Béarn
in France)

PYRENEES

NAVARRE

Barcelona

ARAGON

Ebro

Saragossa

BALEARIC IS.

(& Sicily
& Sardinia
to Aragon)

Valencia

MADRID

Madrid

CASTILE

Toledo

Tagus

Guadiana

Badajoz

Cordova

GRANADA

Malaga

ARAGON

PORTUGAL

Douro

Oporto

Lisbon

Seville

Cadiz

J.F.H.

end of 14ᵗʰ cent.

108

SPAIN:
(iii) AT END OF
14th CENTURY

DURING the thirteenth and fourteenth centuries the Arabs were pushed steadily southward until they held only the province of Granada in the extreme south of the peninsula. Castile now reached to the sea on the south and east, as well as on the north and west. Portugal had now extended to its present-day frontiers.[1] Aragon had become a sea-power, having conquered not only the Balearic Islands, but also Sardinia and Sicily.

Granada remained as a subject province until 1492, when the Arabs were finally expelled from Spain, and the union of Aragon and Castile by the marriage of Ferdinand and Isabella completed the unification of the Spanish state as we know it to-day.

[1] From 1580 to 1640 it was subject to Spain, the whole peninsula during that period being thus under one rule.

MAP 55

SWEDEN

NORWAY

DENMARK

D. of POLAND

Elbe

THE EMPIRE

Rhine

Friesland

LORRAINE

Flanders

NORMANDY

NORTH SEA

Shetlands

Orkneys

SCOTLAND

Lothian

Northumbria

E. Anglia

Mercia

Essex

Wessex

N. Wales

W. Wales

IRELAND

J.F.H.

A.D. 1016-1035

ENGLAND:
(i) THE DANISH
CONQUERORS

THE ENGLAND of the modern world is the result of a long series of invasions from north-east and south. The last two of these invasions were both made by the Norsemen; first by the Danes, who, as pagan pirates, ravaged the coasts of Britain and Ireland and in the ninth century conquered all eastern England between Tees and Thames; and secondly by the Normans from France—Norsemen who had conquered and settled that part of northern France which came to be called Normandy, and who had accepted Christianity.

In the eleventh century, just before the Norman conquest of England, an able Danish ruler, Canute, made the whole of this country part of his Scandinavian empire, which included Norway as well as Denmark. The influence of England and its culture on the Danes was probably much greater than that of the Danes on England; but it is important to remember that at this period England was a subordinate part of a continental empire.

MAP 56

SCOTLAND

IRELAND

York

Wales

ENGLAND

London

NORMANDY

Brittany

MAINE

ANJOU

Paris

F R A N C E

Loire

Poitou

AQUITAINE

Dominions of
Angevin kings

J.F.H.

Gascony

Toulouse

NAVARRE

Roussillon

ARAGON

THE EMPIRE

12th century

ENGLAND:
(ii) THE NORMAN
CONQUERORS

THE NORSEMEN, who, after two centuries' settlement in France, crossed the Channel and conquered England (thirty years after Canute's death), had already proved their capacity, not only for military achievement, but in the arts of government and social organisation. But, for a century and a half after the Conquest, England formed only a part, and at certain periods by no means the most important part, of their dominions. The French domains of the Angevin kings—Henry II, Richard I, John—exceeded in extent their English possessions. The beginning of the long struggle between English and French monarchs for dominance in France was when, in John's reign, the French took Normandy and Anjou. Two and a half centuries more elapsed ere the English crown lost Aquitaine, and yet another century before the last English possession in France, Calais, was lost in 1552.

It was Henry II who, in the middle of the twelfth century, gained the sanction of the Pope for the first English attempt to conquer Ireland.

MAP 57

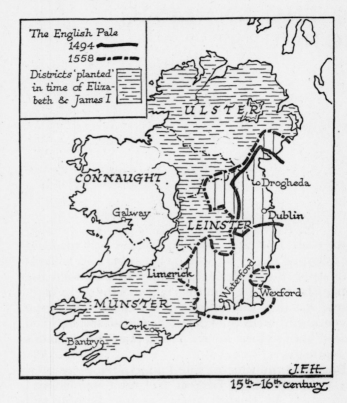

The English Pale
1494 ▬▬▬
1558 ▬ ▬ ▬
Districts 'planted'
in time of Eliza-
beth & James I

ULSTER

CONNAUGHT

Galway

LEINSTER

Drogheda

Dublin

Limerick

Waterford

Wexford

MUNSTER

Cork

Bantry

J.F.H.

15ᵗʰ–16ᵗʰ century

IRELAND:
(i) THE ENGLISH
CONQUEST

THE FIRST ENGLISH attack on Ireland was made in
the reign of Henry II, who (1154) obtained a Papal
Bull—from the only Englishman who ever held the
Papal Chair—giving him the right to subdue Ireland.
There was thus the sanctity of a Crusade about the
adventure; though the Irish were Christians, and had
indeed been so for a longer period than either Saxons
or Normans.

But until the time of the Tudors, despite numer-
ous wars, the larger part of Ireland remained un-
conquered. An English "pale" around Dublin and
Drogheda was held by the invaders. Henry VIII,
Elizabeth and James I conquered further large areas,
"planting" them with Scottish and English settlers.
Practically the whole of Ulster was, in 1611, settled by
Presbyterian Scots, the Irish being driven westward.

MAP 58

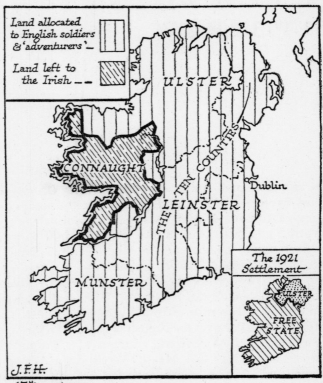

Land allocated to English soldiers & 'adventurers'—

Land left to the Irish — —

ULSTER

CONNAUGHT

THE TEN COUNTIES

LEINSTER

Dublin

MUNSTER

The 1921 Settlement

ULSTER

FREE STATE

J.F.H.

17th century

IRELAND:
(ii) CROMWELL'S
SETTLEMENT

In the middle of the seventeenth century Cromwell, Lord Protector of England, undertook the final conquest of Ireland. England was now consolidating her position as a sea-power; and a possible rival so near to her western coasts could not be tolerated. By the Act of Satisfaction (1653) Ireland was divided into two parts, one consisting of part of the province of Connaught, which was left to the Irish—with the exception of a four mile strip along the coast-line—the other being allocated to English settlers and soldiers. Ten counties—Waterford, Limerick, Tipperary, Queens and Kings Co., Meath, Westmeath, Armagh, Down and Antrim—were awarded to the "adventurers" and soldiers who had been engaged in the work of conquest.

The settlement of 1921 was the final reversal of Cromwell's decree; though the Scottish Presbyterian part of Ulster still insisted on separation from the Irish Free State.

MAP 59

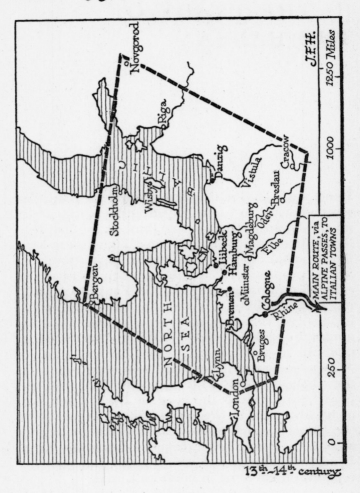

NORTH SEA

Novgorod

Riga

Danzig

Vistula

Cracow

Breslau

Oder

Elbe

Stockholm

Wisby

Bergen

Lübeck

Hamburg

Münster Magdeburg

Bremen

Cologne

Rhine

Lynn

London

Bruges

MAIN ROUTE, via
ALPINE PASSES, TO
ITALIAN TOWNS

J.F.H.

1250 Miles

1000

250

0

13th–14th century

118

THE BALTIC:
(i) THE TRADE OF
THE HANSA

FROM THE eighth to the tenth century the Baltic Sea
was dominated by the Norse adventurers who sailed
up the Russian rivers and established the first civilised settlements in eastern Europe. From its southern
shores had already marched and sailed westward the
Saxon peoples who conquered Britain. By the
eleventh century the Holy Roman Empire was pushing its eastern frontier towards the river Oder; and a
little later Teutonic Knights were waging a crusade
against the pagan Prussians, finally holding the whole
coast-line up to Riga.

Then in the thirteenth and fourteenth centuries
the whole southern Baltic area, together with the
North Sea as far as the coast of Britain, was economically united by the great trading organisation of
the Hansa League, with its most important centres
at Lubeck, Danzig and Cologne. London and Kings
Lynn were thus linked with Sweden and with the
republic of Novgorod in northern Russia.

MAP 60

Swedish territory

NORWAY

SWEDEN

FINLAND

Stockholm

INGRIA

ESTONIA

Novgorod

LIVONIA

Windau

D.vina.

Memel

DENMARK

W. POMERANIA

Pillau

Danzig

Hamburg

Oder

Vistula

POLAND

Bremen

Elbe

THE EMPIRE

J.F.H.

17th century

120

THE BALTIC:
(ii) A SWEDISH LAKE

In the sixteenth and seventeenth centuries, as the trade of the Hansa declined with the opening of the new world ocean routes, and the wars of religion raged in northern Europe, Sweden, a Protestant Power, made herself mistress of the Baltic. During the reigns of Gustavus Adolphus (1611–32) and his daughter, Christina (1632–54), wars with Poland, Russia and Denmark gave Sweden possession of large territories on the eastern and southern shores. In the middle of the seventeenth century, besides these actual possessions, she had the right to levy tolls on the ports of Windau, Memel, Pillau, and Danzig.

MAP 6I

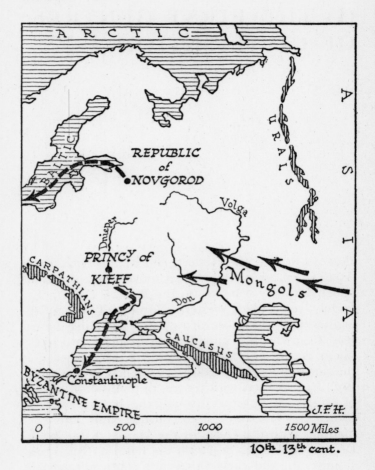

RUSSIA:
(i) THE FIRST CENTRES
OF CIVILISATION

Across Russia, the western end of the great Euro-Asian plain, swept those great movements of peoples which ultimately broke up the Roman Empire and largely re-peopled Western Europe. Then in the eighth and ninth centuries came invading Norsemen from the Baltic, and these founded two settlements destined to last for centuries, at Novgorod in the north, and at Kiev on the Dnieper, in the south. Both had contacts with centres of civilisation outside of Russia: Novgorod, through the Baltic, with the countries of north-western Europe, and later (*cf.* map 59) with the Cities of the Hansa; Kiev, by the Dnieper and Black Sea, with the Eastern (Byzantine) Empire. In the early thirteenth century came the great invasion of the Mongols under Jenghis Khan.

MAP 62

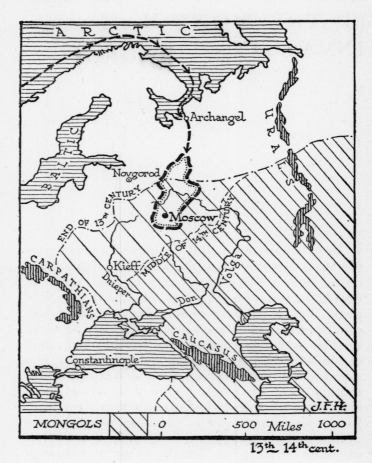

ARCTIC

BALTIC

URALS

oArchangel

Novgorod

END OF 13ᵗʰ CENTURY

•Moscow

14ᵗʰ CENTURY

MIDDLE OF

oKieff

Dnieper

CARPATHIANS

Don

Volga

CAUCASUS

Constantinople

J.F.H.

MONGOLS

0   500   Miles   1000

13ᵗʰ ⁑ 14ᵗʰ cent.

124

RUSSIA:
(ii) THE MONGOL
INVASION

UNDER JENGHIS KHAN and his immediate successors
the Mongols swept across the whole of southern and
central Russia. The principality of Kiev was conquered, as was a large part of the republic of Novgorod. But less than a century later the wave had
considerably receded, and a new principality, Moscow,
situated in the very midst of the central plains, took
the lead in driving the Mongols further back. The
expansion of the new state was rapid. In the fifteenth
century it absorbed the republic of Novgorod and—
the age-old struggle of the plain-dwellers for a coast-
line—pushed northward to Archangel and the shores
of the Arctic. It was from this direction that English
merchant adventurers, searching for a north-east
passage in Tudor times, established contact with
Muscovy.

MAP 63

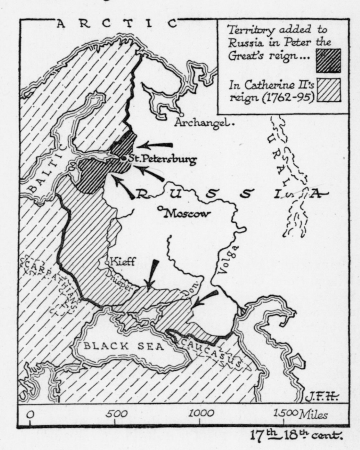

A R C T I C

Territory added to
Russia in Peter the
Great's reign...

In Catherine II's
reign (1762-95)

Archangel.

St. Petersburg

R U S S I A

BALTIC

Moscow

U R A L S

Kieff

Dnieper

Don

Volga

CARPATHIANS

BLACK SEA

CAUCASUS

J.F.H.

0   500   1000   1500 Miles

17th-18th cent.

126

RUSSIA:
(iii) PRESSURE
TOWARDS
THE SEA

PETER THE GREAT, the strange madman who ruled
Russia from 1689 to 1725, by wresting from the
Swedes a considerable part of their possessions on the
eastern shores of the Baltic, achieved his ambition of
giving his country "a window opening on to Europe."
He made his capital the new city of St. Petersburg.
His successors during the eighteenth century extended
Russian territory still further westward and southward. Poles and Lithuanians were driven back from
the line of the Dnieper; and the northern shores of
the Black Sea were taken from the Turks. At last, the
people of the Great Plain had direct sea communica-
tion with the outer world—though still only through
inland seas, the outlets of which were controlled by
other states.

MAP 64

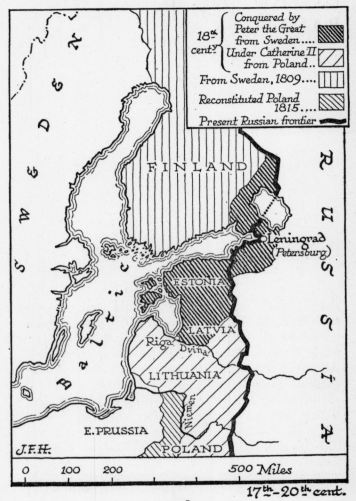

Conquered by
Peter the Great
from Sweden....
Under Catherine II
from Poland..
18th
cent?
From Sweden, 1809....
Reconstituted Poland
1815....
Present Russian frontier

SWEDEN

FINLAND

RUSSIA

Leningrad
(Petersburg)

ESTONIA

Baltic

LATVIA

Riga   Dvina

LITHUANIA

Niemen

E. PRUSSIA

POLAND

J.F.H.

0   100   200   500 Miles

17th–20th cent.

RUSSIA:
(iv) RUSSIA AND THE BALTIC

As we have already seen, Russia secured a Baltic coast-line during the seventeenth and eighteenth centuries. A still further extension of territory was made in 1809, when Finland was taken from Sweden. All these territories, excepting only the area at the extreme end of the Gulf of Finland, immediately surrounding Petersburg (Leningrad), were lost to Russia by the settlement after the Great War of 1914–18. Finland became an independent state, as did the old Baltic provinces of Estonia, Latvia and Lithuania.

MAP 65

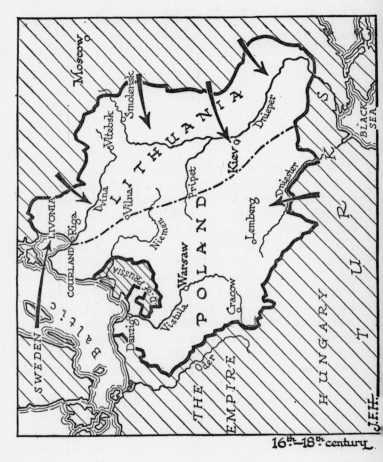

MAP 65

Moscow

Smolensk

LITHUANIA

Vitebsk

Dnieper

Dvina

Vilna

Kiev

BLACK SEA

U   K   R   A   I   N   E

Riga

LIVONIA

Pripet

Dniester

COURLAND

Niemen

POLAND

Lemberg

SWEDEN

Baltic

Danzig

Dl. of PRUSSIA

Warsaw

Vistula

Cracow

Oder

THE

EMPIRE

HUNGARY

16th–18th century

J.F.H.

130

POLAND:
(i) EXPANSION AND
COUNTER-ATTACK

THE NUCLEUS of the Kingdom of Poland was that
section of the great northern plain of Europe lying
between the rivers Oder and Vistula. The latter river
gave the Poles access to the Baltic Sea; but they held
only a small strip of coast-line, the land originally
inhabited by the Prussians and conquered by the
Teutonic Order, blocking their further access to the
sea. To the north-east of the Polish lands lay the
Duchy of Lithuania, and in 1569 the Union of
Lublin amalgamated the two countries. At this time
(see map) the Polish-Lithuanian frontiers almost
covered the territory between the Empire on the
west and Moscow on the east; while north and south
they stretched from the Baltic nearly to the Black
Sea.

The seventeenth century saw the beginning of a
contraction of these extensive territories. On the
north the Swedes took Livonia; from the east the
Russians pressed towards Kiev and the Dnieper; while
on the south the Poles had to resist Turkish aggression.

MAP 66

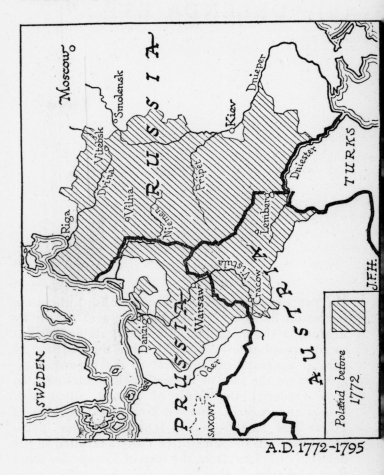

A.D. 1772-1795

Poland before 1772

J.F.H.

132

POLAND:
(ii) "DELENDA EST..."

By the eighteenth century the Polish-Lithuanian territories had been reduced to the area shown by shading in the map. The lack of any stable internal social organisation combined with the absence of natural geographical frontiers to make further aggression on the part of Poland's neighbours easy. In 1772 Frederick the Great of Prussia, allied with Russia and Austria, annexed large outlying areas of Polish land. In 1793 and in 1795 two further "partitions" between the same three Powers resulted in the deletion of Poland from the map of Europe for more than a century.

MAP 67

THE EMPIRE

Kieff

Danube

Vienna

CARPATHIANS

Dniester

HUNGARY

TRANS-SYLVANIA

Drave

WALLACHIA

Venice

Save

Danube

(V.)

SERBIA

Nish

Kossovo

BULGARIA

Uskub

Constantinople

EASTERN EMPIRE

K. of NAPLES

OTTOMAN TURKS

(V.)

(V.)

(V.)

CRETE (V.)

Servian kingdom of Stephen Dushan

Venetian possessions (V.)

J.F.H.

| 0 | 250 | Miles | 500 |

14th century

134

THE BALKANS:
(i) THE COMING OF
THE OTTOMAN
TURKS

WHEN IN THE middle of the fourteenth century the
Ottoman Turks crossed the Dardanelles into Europe,
the dominant power in the Balkan peninsula was the
Serbian kingdom of Stephan Dushan (1331–55). The
Eastern Empire, still with its capital at Constantinople, was now reduced to a mere strip of territory
to the north of the Aegean Sea and the Sea of Mar-
mora. Bulgaria, which some four centuries earlier
had been an empire extending right across the
peninsula from Adriatic to Black Sea, was now a
vassal state of the Serbs. The Venetians (*cf*. map 48)
held most of the harbours of the Adriatic and the
most important Aegean islands, including Crete.
Moving north-westwards, the Turks crushed the Serbs
at the Battle of Kossovo (1389) and before the end of
the century they dominated the whole peninsula up
to the Danube, excluding only the Constantinople
corner. Constantinople itself fell half a century later,
in 1453 (*cf*. map 13).

MAP 68

THE EMPIRE

POLAND

Kieff

Danube

Vienna

CARPATHIAN

Dniester

Budapest

HUNGARY

TRANS-SYLVANIA

MOLDAVIA

JEDISAN

Venice

Drave

Save

OTTOMAN

WALLACHIA

SERVIA

Danube

Adriatic

MONTENEGRO

BULGARIA

Constantinople

Otranto

RUMELIA

EMPIRE

TWO SICILIES

(V.)

Lepanto

(V.)

Aegean

CRETE

(V.)

J.F.H.

(V. = Venetian possessions)

| 0 | 250 | 500 | 750 Miles |

late 16th century

136

THE BALKANS:
(ii) THE OTTOMAN
EMPIRE AT ITS
GREATEST EXTENT

Before the end of the sixteenth century the Turks
had extended their Empire in Europe northwards,
across Hungary, to the Carpathians and north-east-
wards along the shores of the Black Sea. In 1529 they
laid siege to Vienna itself. Earlier they had crossed
the Adriatic and taken Otranto, in southern Italy;
but their naval power was never strongly developed,
and in 1571 the Venetians, with the help of Philip of
Spain, destroyed any possibility of Turkish dominance in the Adriatic by the victory of Lepanto.

All the former kingdoms or principalities of the
Balkan peninsula were now possessions or tributary
states of the Sultan; with the exception of the small
mountain land of Montenegro.

MAP 69

THE BALKANS:
(iii) THE RE-BIRTH OF
THE BALKAN NATIONS

At the beginning of the nineteenth century the Ottoman Empire in Europe still included practically the whole of the Balkan peninsula. Austria (*cf.* map 33) had regained the whole of Hungary. Russia, by a series of wars, had cleared the Turks from the northern shores of the Black Sea, in 1812 taking Bessarabia.

Then, early in the century, the subject nationalities one by one began to revolt. First Serbia, then the Danubian principalities of Wallachia and Moldavia, and then Greece, rose in rebellion. By the end of the century there were four independent states: Serbia, at first, from 1817, a principality, in 1882 a kingdom; Rumania, formed from the unification of Moldavia and Wallachia (1861), and proclaimed a kingdom in 1881; Greece, whose first war of independence was fought in 1821–9, constituted a kingdom in 1863; and Bulgaria, which as a result of Russian intervention and the Russo-Turkish War (1878) was by the Treaty of Berlin made autonomous, although complete independence was not finally achieved until 1908, after the Young Turk Revolution. The province of Bosnia was first (1878) occupied, and later annexed by Austria (*cf.* map 38).

MAP 70

Turkish territory taken by Servia

by Bulgaria

by Greece ..

by Montenegro

New state of Albania ...

Bulgarian territory ceded to Rumania ..

20th century

140

THE BALKANS:
(iv) THE BALKAN WARS, 1912 — 13

TAKING ADVANTAGE of the Turkish defeat by Italy in
1911–12—a war fought mainly for the Turkish territories in North Africa—the Balkan states declared war
on the Ottoman Empire in October 1912, and in the
course of a two months' campaign were in possession
of practically all that remained of Turkish territory
in Europe. Balkan representatives, together with
representatives of the Great Powers, met Turkish
delegates at a conference in London. The claims of
the various victor states proved irreconcilable and
the second Balkan war, between the victorious states,
followed. In this war Bulgaria was heavily defeated
by her late allies, Serbia, Greece and Rumania. The
map shows the division of Turkish territory between
the five states. As a result of Austrian and Italian
opposition to Serbia's claims for an Adriatic coast-
line, the Great Powers decreed that a new state,
Albania, be set up.

INDEX

The numbers are those of maps, but in all cases the entries refer to the letterpress accompanying the map as well as to the map itself.

i

ii

A NOTE ON THE TYPE
IN WHICH THIS BOOK IS SET

The text of this book was set on the linotype in Baskerville. The punches for this face were cut under the supervision of George W. Jones, an eminent English printer. Linotype Baskerville is a facsimile cutting from type cast from the original matrices of a face designed by John Baskerville. The original face was the forerunner of the "modern" group of type faces. ¶ John Baskerville (1706-75), of Birmingham, England, a writing-master, with a special renown for cutting inscriptions in stone, began experimenting about 1750 with punch-cutting and making typographical material. It was not until 1757 that he published his first work, a Virgil in royal quarto, with great-primer letters. This was followed by his famous editions of Milton, the Bible, the Book of Common Prayer, and several Latin classic authors. His types, at first criticized as unnecessarily slender, delicate, and feminine, in time were recognized as both distinct and elegant, and both his types and his printing were greatly admired. Printers, however, preferred the stronger types of Caslon, and Baskerville before his death repented of having attempted the business of printing. For four years after his death his widow continued to conduct his business. She then sold all his punches and matrices to the Société Littéraire-typographique, which used some of the types for the sumptuous Kehl edition of Vol-
—*taire's works in seventy volumes.*—

COMPOSED, PRINTED, AND BOUND BY

H. WOLFF ESTATE, NEW YORK.

THE PAPER WAS MADE BY

S. D. WARREN CO.

BOSTON